P9-AEZ-597

Hydrogen: Production and Marketing

W. Novis Smith, EDITOR
General Electric Company

Joseph G. Santangelo, EDITOR
Air Products & Chemicals, Inc.

Based on a symposium

sponsored by the Division

of Industrial and Engineering

Chemistry at the ACS/CSJ

Chemical Congress, Honolulu,

Hawaii, April 2–6, 1979.

ACS SYMPOSIUM SERIES **116**

AMERICAN CHEMICAL SOCIETY
WASHINGTON, D. C.   1980

SEP/AE
CHEM

6355-0672 ✓

XD 80
2552
CHEM

Library of Congress CIP Data

Hydrogen, production and marketing.
  (ACS symposium series; 116 ISSN 0097–6156)

  Includes bibliographies and index.

  1. Hydrogen—Congresses.
  I. Smith, William Novis, 1937–   . II. Santangelo,
Joseph G., 1930–   . III. American Chemical Society.
Division of Industrial and Engineering Chemistry. IV.
ACS/CSJ Chemical Congress, Honolulu, 1979. V.
Series: American Chemical Society. ACS symposium
series; 116.

TP245.H9H9   338.4'766581   80–386
ISBN 0–8412–0522–1   ACSMC8 116 1–416 1980

Copyright © 1980

American Chemical Society

All Rights Reserved. The appearance of the code at the bottom of the first page of each article in this volume indicates the copyright owner's consent that reprographic copies of the article may be made for personal or internal use or for the personal or internal use of specific clients. This consent is given on the condition, however, that the copier pay the stated per copy fee through the Copyright Clearance Center, Inc. for copying beyond that permitted by Sections 107 or 108 of the U.S. Copyright Law. This consent does not extend to copying or transmission by any means—graphic or electronic—for any other purpose, such as for general distribution, for advertising or promotional purposes, for creating new collective works, for resale, or for information storage and retrieval systems.

The citation of trade names and/or names of manufacturers in this publication is not to be construed as an endorsement or as approval by ACS of the commercial products or services referenced herein; nor should the mere reference herein to any drawing, specification, chemical process, or other data be regarded as a license or as a conveyance of any right or permission, to the holder, reader, or any other person or corporation, to manufacture, reproduce, use, or sell any patented invention or copyrighted work that may in any way be related thereto.

PRINTED IN THE UNITED STATES OF AMERICA

SD
5/15/80
JW

TP 245
H9H9
CHEM

ACS Symposium Series

M. Joan Comstock, *Series Editor*

Advisory Board

David L. Allara

Kenneth B. Bischoff

Donald G. Crosby

Donald D. Dollberg

Robert E. Feeney

Jack Halpern

Brian M. Harney

Robert A. Hofstader

W. Jeffrey Howe

James D. Idol, Jr.

James P. Lodge

Leon Petrakis

F. Sherwood Rowland

Alan C. Sartorelli

Raymond B. Seymour

Gunter Zweig

05485

FOREWORD

The ACS SYMPOSIUM SERIES was founded in 1974 to provide a medium for publishing symposia quickly in book form. The format of the Series parallels that of the continuing ADVANCES IN CHEMISTRY SERIES except that in order to save time the papers are not typeset but are reproduced as they are submitted by the authors in camera-ready form. Papers are reviewed under the supervision of the Editors with the assistance of the Series Advisory Board and are selected to maintain the integrity of the symposia; however, verbatim reproductions of previously published papers are not accepted. Both reviews and reports of research are acceptable since symposia may embrace both types of presentation.

CONTENTS

PREFACE

There has been a significant amount of literature published and a number of symposia given concerning hydrogen. However, much of this effort and information has been without the involvement of that industrial sector which is involved commercially in the production, utilization, and marketing of hydrogen. The continued and active involvement of the industrial sector in the discussions and meetings concerning the economics and technology of hydrogen production and utilization is necessary to place the current and future role of hydrogen in perspective, and to underscore the technical requirements and constraints for an even greater role for hydrogen in new uses.

To this end, and as part of the ACS/CSJ Chemical Congress in Honolulu, Hawaii, the symposium, "Production and Marketing of Hydrogen: Current and Future" was included as part of its program activities. This symposium focused on both the shorter-term economics and production of hydrogen and the potential longer-term uses of hydrogen. The speakers who were invited to participate represented many of the leading firms currently involved in hydrogen production and utilization as well as informed speakers from the more involved and knowledgeable non-commercial organizations in the hydrogen technology area. A number of longer-term technologies for hydrogen production also were included to point out some of the possible technology changes that could occur.

In an effort to provide a common base for the economics of this symposium, a common list of cost assumptions was used as the basis for most of the chapters. (This list is included as the Appendix.) This volume presents this informative symposium to the larger audience who wishes to keep current on the technology, economics, and production relating to hydrogen.

We wish to acknowledge and thank Linda Anthony for organizing and handling the communications regarding the symposium and this subsequent book.

JOSEPH G. SANTANGELO
Air Products & Chemicals, Inc.
P. O. Box 538
Allentown, PA 18105

July 12, 1979

W. NOVIS SMITH
General Electric Company
3198 Chestnut St.
Philadelphia, PA 19101

CONTRIBUTING AUTHORS

JACK ARORA is Supervisor of Process Economics at the Institute of Gas Technology. He contributes to process economics studies in the evaluation of new technologies for conversion of coal, oil, shale, and biomass to clean and useful fuels.

CHARLES R. BAKER, who holds the position of Engineering Associate with the Linde Division of Union Carbide Corporation, has over 30 years of experience covering a broad range of cryogenic technology and industrial gas processing. He earned his BS and MS degrees in chemical engineering at Pennsylvania State University and is a member of the American Institute of Chemical Engineers and the International Association for Hydrogen Energy.

MORRIS BELLER received his BChE from the Polytechnic Institute of Brooklyn in 1955 and his MS from the State University of New York at Stony Brook in 1972. He was employed by the Linde Division of the Union Carbide Corporation from 1956–1959 and by the Cornell Aeronautical Laboratory from 1959–1962. He has been employed by Brookhaven National Laboratory since 1962. Mr. Beller is involved with chemical plant design, start-up, and evaluation, chemical process economic assessment, and technology assessment of energy-related processes and equipment. Present interests are in the areas of potential impacts of energy technologies and policies and in energy contingency planning.

BEVERLY J. BERGER is Chairman of the U.S. DOE Hydrogen Energy Coordinating Committee. She came to DOE as Program Manager for the Chemical Storage Systems program from the Energy Resource Planning Group at Lawrence Livermore Laboratory. Dr. Berger, a population geneticist, is now Policy and Planning Manager for the Biomass Energy Systems Branch within the U.S. DOE.

L. J. BUIVIDAS is employed by Pullman Kellogg of Houston, TX as an Inorganic Chemical Technology Coordinator. He is responsible for developing new technology in order to improve the marketing position of Pullman Kellogg in the chemical industry. He received a BS in Chemical Engineering from Tri-State University in 1942; he also holds a BS in Petroleum Engineering and an MS in Petroleum Refinery Engineering from Tulsa University.

JORGE A. CAMPS has BS and MS degrees in Chemical Engineering from Louisiana State University. He worked for five years with Exxon Corporation at various U.S. and overseas locations. He joined Davy Powergas International in 1974 and is now a Principal Process Engineer of Synthesis Gas Processes. His most recent experience was as the Lead Process Engineer for a 2300-STPD methanol plant for SCT in Saudi Arabia. Mr. Camps is also an adjunct Professor of Chemical Engineering at the University of South Florida in Tampa.

GIOVANNI CAPRIOGLIO is manager of the Materials and Chemistry Department at the General Atomic Company. He received his PhD in Industrial Chemistry from the University of Milan.

HAMPTON CORNEIL has been a senior advisor for Exxon Enterprises Inc., an affiliate responsible for developing diversification opportunities for Exxon. Since 1975, he has been concerned with opportunities in energy conversion and storage. As a part of this effort, he has conducted various studies concerned with future markets for hydrogen.

T. A. CZUPPON is a Process Manager at Pullman Kellogg in Houston, TX where he manages the preparation process design and economic evaluations for chemical facilities, principally in the areas of ammonia and related fertilizers, hydrogen, methanol, and synthesis gas plants. He received his BS from C. W. Post College, Long Island University in 1966 and his MS in Chemical Engineering from New York University in 1968.

JÜRGEN FALBE is Executive Vice President of Ruhrchemie AG, Oberhausen. He received his PhD in Chemistry from the University of Bonn in 1959. His responsibilities include research and development, engineering and application technologies, patents, licensing, documentation, and environmental affairs.

G. H. FARBMAN joined the Advanced Energy Systems Division of Westinghouse Electric Corporation in 1974. He has the responsibility for development and demonstration programs in emerging energy related technologies, including hydrogen production, gas-cooled nuclear reactors, propulsion systems, coal bed methane recovery, and flue-gas treatment systems.

DEREK GREGORY is Vice President, Engineering Research, at the Institute of Gas Technology in Chicago, where he is responsible for programs in Energy Conservation, Energy Conversion and Storage, and in Alternative Energy Systems. He has had many years of direct involvement in hydrogen and fuel cell research projects. Dr. Gregory received his BS and PhD in Physical Chemistry from Southampton University in England.

FRED J. HEINZELMANN is employed by the Exxon Research and Engineering Company in Linden, NJ.

TETSUO IWASAKI obtained his BS in Metallurgical Engineering at Hokkaido University and is a graduate student in the master's course of Metallurgical Engineering at the same university.

HIROMICHI KIUCHI received his BS in Geology and Mineralogy at Hokkaido University in 1963. Since that time he has been a research associate of nonferrous extractive metallurgy in the Department of Metallurgical Engineering at the university.

CALVIN J. KUHRE is a senior engineer with the Shell Oil Company in Houston, TX. His responsibilities include process engineering design for the Shell Gasification Process and other petrochemical–refining processes. He also is involved in various other aspects of licensing Shell gasification technology. He received his BChE from Cornell University and did graduate work at the University of California at Berkeley in chemistry.

KENNETH H. McCORKLE was graduated from the Massachusetts Institute of Technology with a BS and MS in Chemical Engineering. He also received a PhD in Chemical Engineering from the University of

Tennessee. He worked on the processing and fabrication of fresh and irradiated nuclear fuels materials at Oak Ridge National Laboratory and at the General Atomic Company for 20 years. For the past five years he has done conceptual and bench-role process design for hydrogen production processes, principally for thermochemical water-splitting processes.

MANUS McHUGH, III, General Manager, IGD Operations at Air Products and Chemicals, Inc., holds both a BS and MS degree. He has 15 years of experience in sales, engineering, distribution, and operations.

CLYDE McKINLEY has been with Air Products and Chemicals in R&D roles for over 25 years. Much of this period has been in cryogenic technology, basic properties studies, and process development. His personal involvement in the liquid hydrogen program in America provides a unique perspective for his chapter.

A. MEZZINA is employed by Brookhaven National Laboratory in Upton, NY.

R. G. MINET is Chairman of the Board and Chief Executive Officer of KTI Corporation in Pasadena. Prior to that position he served as President of KTI Corporation in Pasadena and was Group Vice President for KTI BV in the Netherlands, prior to establishing the U.S. company. He holds a PhD in Engineering from New York University, an MS from Stevens University, New Jersey and a BChE from City College in New York. He has served as an Executive Officer for several European and U.S. engineering contractors. He is a registered Professional Engineer in five states in the U.S.A. He is an active member of AIChE and has been a member of the American Chemical Society since 1949.

ISAO NAKAMURA obtained his BS in Metallurgical Engineering at Hokkaido University and is a graduate student in the master's course of Metallurgical Engineering at the same university.

PRASAD V. NEVREKAR is an associate chemical engineer with the Institute of Gas Technology. His current responsibilities include analysis and evaluation of data for various coal conversion processes. He earned his BS and MS degrees in Chemical Engineering from the University of Madras (India) and the Illinois Institute of Technology, respectively.

LEONARD J. NUTTALL, a University of Utah graduate, BSME, 1946, has more than 30 years of design, product planning, and applications engineering experience with the General Electric Company. After joining the General Electric Company in 1946, Mr. Nuttall served as a test engineer principally associated with Aeronautical and Aircraft Gas Turbines. Following this, he spent five years as an aerodynamic design and development engineer, and held subsequent assignments in Advanced Product Planning and Applications Engineering principally in Aircraft and Aerospace fields. In 1962, Mr. Nuttall was assigned the responsibility for applications engineering of direct energy conversion products including fuel cell electrical power supplies, electrolysis systems, oxygen concentrators, and related products. Since 1969, he has directed all applications engineering activities for government- and industry-sponsored development programs (principally major aerospace applications), electrolysis life support systems for submarines, and was responsible for extending aerospace-developed technologies to a number of commercial energy- and environmental-related development programs and products.

O. OLESEN is employed by the United Technlogies Corporation in South Windsor, CT.

G. H. PARKER is Manager of Hydrogen Production Programs at Westinghouse Electric Corporation. He is responsible for developing the Westinghouse thermochemical process. He has been involved in the design, development, and testing of a variety of high technology systems for more than 20 years and formerly was Manager of System Design and Analysis at Westinghouse.

M. E. D. RAYMONT is Director of R&D at the Sulphur Development Institute of Canada in Calgary, Alberta.

CALVIN L. REED is a staff engineer with the Shell Oil Company in Houston, TX. His responsibilities include process engineering and sales work with respect to the licensing of the Shell Gasification Process. Mr. Reed received his BA from Albion College and his BS in Chemical Engineering from the University of Michigan.

FRANK J. SALZANO has been employed as a chemical engineer at Brookhaven National Laboratory since 1957. He is currently Head of the Energy Storage and Conversion Division where he is responsible for programs in hydrogen technology, fuel cells, and conservation. He is involved in process chemical engineering, surface chemistry, solid-state chemistry, the chemistry of liquid metals, fused-salt electrochemistry, and engineering systems studies.

GARY D. SANDROCK was formally trained in metallurgy at the Illinois Institute of Technology and Case Western Reserve University, receiving his BS, MS, and PhD in 1962, 1965, and 1971, respectively. From 1962–1969 he was a Research Metallurgist at the NASA Lewis Research Center. Since 1971 he has been a Senior Project Engineer at Inco's R&D Center at Sterling Forest. There his main interest has been in hydrogen and rechargeable metal hydrides. He originated Inco's extensive effort in this area. He has 32 publications and 12 patents in the fields of high-temperature alloys, phase transformations, and metal hydrides.

WARREN G. SCHLINGER received his PhD in Chemical and Mechanical Engineering from the California Institute of Technology in 1949. He currently is manager of the Montebello Research Laboratory of Texaco, Inc.

R. SHARP is employed by the General Atomic Company in San Diego, CA.

JAMES L. SKINNER received a BS in Chemical Engineering from the University of Illinois and an MS in Chemical Engineering from the University of Michigan. He joined Atlantic Richfield R&D in 1964 and worked in Production Research until 1969, when he joined the Synthetic Crude and Minerals Research Group in ARCO. Presently he is a Director in the Synfuels, Coal, and Minerals Research Group.

E. SNAPE obtained his BS and PhD degrees in Metallurgical Engineering from the University of Leeds in 1965. During his research career with Inco, he received numerous awards for his work on stress corrosion and hydrogen embrittlement, and became an international authority on this subject. In 1974, Dr. Snape pioneered the development of a waste reclamation process which now is being applied to the recovery of scarce domestic mineral resources. He then turned his attention toward the alternate energy field and managed a research and development program which led to the development of new types of reversible metal hydrides and unique systems for utilizing these hydrides. With Inco's support, Dr. Snape established a subsidiary company to manufacture and market products and systems involving solar and hydrogen technologies. Dr. Snape is a member of the International Association for Hydrogen Energy and is active in a number of government- and industry-sponsored hydrogen energy programs.

RUDOLF SPECKS studied at the Technical University of Aachen in the Department of Mining and Metallurgy 1949–1954. From 1955 until 1958 he was employed by the mining office and in 1959 he was promoted to senior engineer and managed a coal mine until 1970 when he became general manager of the R&D Department of Ruhrkohle AG. Since 1975 he has been Chairman of the Board of Management of GVV, the Ruhrkohle subsidiary for coal conversion.

SUPRAMANIAM SRINIVASAN received his BS from the University of Ceylon in 1955 and his PhD in Physical Chemistry from the University of Pennsylvania in 1963. He currently is employed as a chemist by Brookhaven National Laboratory in Upton, NY.

GERALD STRICKLAND was graduated as a Chemical Engineer for Cleveland State University in 1943 and has had additional training in mechanical engineering. He has worked at Oak Ridge National Laboratory on fission product separation and on process equipment design. In 1947 he joined Brookhaven National Laboratory and has worked on laboratory and equipment design, nonaqueous fuel reprocessing, fission product disposal in phosphate glass, liquid–metal heat transfer research, and metal hydride technology.

JAMES H. SWISHER received his BS in Metallurgical Engineering from the Carnegie Institute of Technology in 1959 and his PhD in Metallurgy and Materials Science from Carnegie–Mellon University in 1963. He served as a First Lieutenant in the U.S. Army Corps of Engineers from 1963–1965. He currently is employed as Assistant Director of Physical Storage Systems with the DOE Division of Energy Storage Systems.

TOKIAKI TANAKA is Professor of Nonferrous Extractive Metallurgy in the Department of Metallurgical Engineering at Hokkaido University where he received his PhD in Engineering in 1962.

CONSTANTINE TSAROS is Manager of Process Economics in the Process Research Division at the Institute of Gas Technology. He has been concerned for several years with development of cost estimates for large-scale process plants, including projections for new technology plants for the conversion of coal into gaseous fuels.

DAVID McG. TURNBULL is a graduate of the University of Canterbury, N.Z. with bachelor's degrees in Chemistry (1970) and in Chemical Engineering (1972). Mr. Turnbull joined the methanol group in the Process Department of Davy International Oil & Chemicals Ltd. in London in 1973. Since then he has worked on a variety of methanol and ammonia projects. In 1977 he transferred to Davy Powergas International in their Lakeland, FL office where he was recently Lead Process Engineer on the design of two 950 STD methanol plants being built by Davy for Borden Chemical in Louisiana.

INTRODUCTION

I have been a member of the ACS since I was a student at Washington State University. In fact, Gardner Stacy was my professor for Organic Qualitative Analysis. Considering the fires I started during that course, I didn't expect that he would invite me anyplace—but the passage of time does blur bad memories. Anyway, I think I've done better in politics than I would have in chemistry.

Of course, there are those who would accuse me of starting fires today in the political world. In that respect, may I express my congratulations to you for the responsible leadership the ACS is offering the people of this country in their consideration of technical matters, and the constructive assistance ACS is providing to the Congress, the Administration, and the press, and in general to the government in its attempts to respond to problems and issues involving technical and scientific subjects.

I am—and I hope you are—proud of Gardner Stacy's public lectures and his leadership in helping to establish responsible and realistic energy policies for our nation. I hope he and other leaders of the ACS will continue such activities. This is public service at its best, and I hope the Internal Revenue Service recognizes it as such.

I'm also proud of and pleased with the ACS support of the Science Fellowship Program, managed by the American Association for the Advancement of Science; I hope we can increase that support. I hope that Members of Congress will, in the immediate future, take greater advantage of the ACS offer to provide technical assistance on issues and questions appearing before the Congress. Such assistance is definitely needed. Members of Congress, for instance, often refer to hydrogen as an energy source, as in: ". . . we must rely on coal, conservation, and hydrogen and solar energy . . ." And some of our constituents confuse Members of Congress with angry letters complaining about the ". . . conspiracy to prevent the mining of hydrogen from the oceans . . ." or about our failure to ". . . drill hydrogen gas wells. . . ." We also receive suggestions such as that hydrogen and natural gas should be ". . . diluted with helium to make it safe . . ." We also, of course, receive mail telling us of perpetual motion machines. These are samples of sincere correspondence that comes from concerned constituents, and it must be treated as such.

They are, of course, somewhat easier to handle than bloopers in speeches and public statements by Members of Congress. Several years ago a Congressman asked me to read his statement on the problem of nitrogenation of the water of the Columbia River as it falls over the face of high dams. The statement ended with a stern warning that if the Corps of Engineers couldn't straighten out the situation, then he (the Congressman) would take action ". . . prohibiting nitrogen altogether . . ."

This provides some insight into the fact that hydrogen, nitrogen, and methane; uranium and coal; nitrites in food, DDT, saccharine, and cyclamates, etc. have political as well as chemical and physical properties.

In the world of Washington, D.C. it may at times be difficult to distinguish the real from the imaginary; the physical and chemical from the political. Part of this arises from the campaign of alienation that is being carried out systematically against the scientific and technological communities by those who want us to live in a world of make-believe, and to prevent the solution of our real-life problems by offering unrealistic, unscientific solutions to the problems that face our nation. I recently was attacked bitterly in a book by a solar energy promoter because, in establishing the solar energy heating and cooling program, I insisted that the National Bureau of Standards be required to set performance standards before federal grants could be awarded for these projects.

Recent movies and so-called "documentaries" accentuate this phenomenon. In the perverted logic of such antiscientific "new-think," anyone who is a qualified physical scientist is disqualified automatically from policy-making in his or her discipline. A true scientist thus is considered to be prejudiced because he or she insists on physical truths. In such a bizarre Catch-22 world, only the scientifically unqualified are politically qualified to make policy decisions on matters dealing with science and technology.

These attitudes attempt to institute a new class—a new elite—a new type of reporter, a new activist, a new expert, qualified by a lack of scientific education, who would make policy and control news, attitudes, and societal action on issues dealing with scientific matters on anything but scientific, engineering, and economic facts. This is, if you haven't noticed it, a growing, real, and frightening phenomenon. Thus, we have those who insist on defining hydrogen as an energy source along with, for instance, gasohol. This is part of an escape mechanism which presumes that we can solve our twentieth-century energy problems with conservation, solar energy, wind, bioconversion, geothermal energy, appropriate technology, and low-head hydroelectricity. Its practitioners presume (if they think at all) that all of us will be better off with a lower standard of living. Of course, these are all valuable technologies. They will contribute to our future energy supply. I have initiated legislation to create and fund aggressive research, development, and demonstration programs which I hope and expect will lead to their early commercialization; but all together they cannot solve our energy problems in this century. Obviously, hydrogen falls into the same category. I believe that we will, during the 1980's, develop technologies for large-scale, commercially competitive hydrogen production; and that its use will become widespread during the 1990's in the following areas:

- as a portable fuel for surface and air transportation;
- for intermediate electric production from fuel cells;
- for a new generation of cryogenic superconductors with an increased potential for cheaper electric energy transmission and large-scale storage of electricity;
- for cheaper superconducting magnets for such uses as new fusion reactors and similar devices; and
- possibly for the manufacture of methane.

All this, of course, requires that economically competitive technologies will be developed. As the costs of oil, gas, and coal continue to increase, the potential economic competitiveness of hydrogen production is becoming more real daily. Accordingly, it is essential that we move forward as rapidly as we can with research and development programs involving hydrogen production and use.

This is an area wherein the House Committee on Science and Technology, and the ACS can interact beneficially. In the House of Representatives Subcommittee on Energy Research and Production which I chair, we have increased the authorization level requested by the Administration for hydrogen research and development from \$18.6M to \$22.6M. This is our figure for the 1980 fiscal year and I hope that the actual funding level will be increased further by subsequent Congressional action. The programs we have authorized include more efficient and cheaper production, transportation, storage, and use of hydrogen and materials studies of metal hydrides, and pipes, pumps, and meters for handling hydrogen gas.

Of course, all this will take time. Any new technology takes 10–30 years to get into significant use. However, we can see that there are several real advantages in a hydrogen economy:

- environmental;
- supply;
- independence;
- efficiency; and
- flexibility

As we develop new hydrogen uses and technologies, we also must help the people of this country to understand the realistic role that hydrogen will play. First, that it is not a source of energy, and that it will not provide a panacea or a magic quick fix to our energy problems; second, that its use will not come quickly or suddenly, but will be phased in slowly as a part of our mix of energy technologies; and third, that hydrogen and hydrogen technologies have much to offer mankind.

As you do this, I hope that you will accept a greater role in your own communities and a more active relationship with the elected public officials who will be making policies concerning hydrogen use, and to make policies with respect to a broad spectrum of other subjects involving scientific questions.

Your leadership in political as well as scientific activity can make a great difference in the future welfare of your country.

U.S. Congress   MIKE McCORMACK

July 12, 1979

OVERVIEW OF HYDROGEN RESEARCH AND DEVELOPMENT

The Economics of Hydrogen Production

D. P. GREGORY, C. L. TSAROS, J. L. ARORA, and P. NEVREKAR

Institute of Gas Technology, 3424 South State Street, Chicago, IL 60616

Hydrogen is used today as a major chemical intermediate, as a reducing gas, and for a wide variety of special applications. As fuel processors begin to deal with "heavier" feedstocks, including coal and shale, hydrogen will play an increasing role in their conversion to clean and useful light hydrocarbon fuels. Consideration is currently being given to the use of hydrogen as a fuel for such applications as automobiles and aircraft, where its cleanliness and light weight offer advantages, and ultimately, as a possible replacement for natural gas and petroleum in the transportation, storage, and delivery of energy derived from non-fossil (nuclear and solar) sources. (In the latter case, hydrogen is considered as a carrier of energy, analogous to electricity, rather than an energy source.)

The cost of producing hydrogen is a dominant factor in all aspects of its present and future use. Present hydrogen manufacturers must consider the changing cost and availability of feedstocks, the rising manpower costs, and the increasing costs of capital. Decisions about the future uses of hydrogen require accurate projections of the cost of manufacturing it by both conventional and still-developing methods.

There are four basically different methods of making hydrogen which must be considered: In the first category, a relatively reactive light hydrocarbon, such as naphtha or natural gas, is reacted with steam, producing hydrogen derived partly from the hydrocarbon itself and partly from the steam. In the second category, heavy, relatively unreactive feedstocks such as coal and residual oil, are partially oxidized and reacted with steam, producing hydrogen. In the third category, water alone is split electrochemically, producing hydrogen and oxygen, the needed electric power coming from one or more of a wide variety of raw energy sources. In the fourth category, water molecules are split in a complex chemical cycle which uses primarily heat as the energy input — all other materials of the cycle being continuously

0-8412-0522-1/80/47-116-003$06.00
© 1980 American Chemical Society

recycled. The heat for such a cycle would be produced from
nuclear or solar energy.

It is true to generalize that the cost of making hydrogen
today is lowest for the first category and highest for the fourth;
however, this does not mean that only the first category is in use
today, although it appears to be the cheapest, nor does it mean
that the sequence of economic priorities will always be in today's
order.

In this paper, we have not attempted to answer the question
of "How much does hydrogen cost?" or of "Which process is the most
economic?" Rather, we have set out, on a parametric basis, a
range of hydrogen costs for a selected variety of production
processes covering all of the categories mentioned. We have shown
the effect of some major variables, such as feedstock price and
by-product credits, where applicable. This allows for an understanding of the criteria underlying any more-detailed analyses of
hydrogen production costs. Additionally, in the scope of this
paper, we have not attempted a detailed process description and
economic analysis of each process considered. We have taken
economic data available in the published literature and converted
the data to a common basis in line with the guidelines (shown in
Table I). These are largely those supplied by the organizers of
this symposium* with the following additions or modifications:

- Interest during construction added: 15% of facilities
  investment
- Land: not included
- Fuel: same as feedstock
- Return on Investment, or cost of capital: 20% before
  taxes = 10% after tax at 50% tax rate.

Table II gives a summary comparison of the major economic elements
for each process. Base plant capacity is 100 million SCF/day of
hydrogen; investment is representative of estimated mid-1979 costs.

Process Descriptions

Hydrogen By Steam Reforming. Historically, steam reforming
is the most widely used and the most economical process for
producing hydrogen from light hydrocarbon gases or from naphthas.
The process is based upon the well-known reaction of steam and
hydrocarbons, which is carried out over a nickel catalyst. With
methane, the reaction is $CH_4 + H_2O = CO + 3H_2$; the reaction is
highly endothermic, and is carried out in a reformer tube furnace
fueled by the feedstock. Exit temperatures range from 1300° to
1600°F. The nickel catalyst is poisoned by sulfur; thus, sulfur
is removed from the feed. Figure 1 is a block flow diagram showing the basic process steps with natural gas as the feed.

Reformer effluent is cooled by waste heat recovery, which
generates reaction steam plus some by-product steam. The carbon

_____

* ACS Spring Meeting: "Production and Marketing of Hydrogen,
   Current and Future," April 5-6, 1979, Honolulu, Hawaii.

Table I.   ECONOMIC EVALUATION GUIDELINES

Project Life: 20 years
Operating Factor: 330 days/year
Capital Investment (mid-1979 basis)
    Facilities Investment (F.I.) = As obtained from literature
                                      and adjusted to mid-1979 basis
    Interest During Construction = 0.15 F.I.
    Startup Expense   = 0.02 F.I.
    Working Capital   = 60 days feedstock utilities
                                        and cash supply
        Total Capital Investment = Sum of the Above

Annual Operating Cost
    Feedstock
        Natural Gas $2/$10^6$ Btu
        Wyoming Coal $20/ton ($1.17/10^6 Btu)
        Montana Coal $17.7/ton ($1.00/10^6 Btu)
        Residual Oil $15/bbl
    Utilities
        Raw Water ($/$10^3$ gal) 0.50
           Power ($/kWhr)   0.030
        Cooling Water   0.05
          ($/$10^3$ gal)
           Fuel ($/$10^6$ Btu) same price as for feedstock

Cash Supply
    Operating Labor $18,000/man-year
    Operating Labor   15% of operating labor
     Supervision
    Maintenance
        Labor   2% of facilities investment
        Supervision   15% of maintenance labor
        Materials   2% of facilities investment
    Administrative and Support 20% of all other labor
     Labor
    Payroll Extras/Fringe   20% of all other labor
     Benefits, etc.
Insurance   2% of facilities investment
General Administrative Expenses 2% of facilities investment
Taxes (Local, State, and   50% of net profit
     Federal) (No investment
     Tax Credit)
Depreciation   Straight line, 5%/year
Return on Investment   20% before taxes (0.2 X C)
 (cost of capital)
Gross Revenue Required   Sum of the Above

By-Product Credits
    Sulfur $10/long ton
    Steam $3/1000 pound
    Power   2¢/kWhr
    Oxygen $10/ton
Net Revenue Required   = Gross Revenue Required — By-Product
                         Credits
Hydrogen Price, $/$10^3$ CF = $\dfrac{\text{Net Annual Revenue Required}}{\text{Annual Hydrogen Production}}$

√ Table II. COMPARISON OF VARIOUS PROCESSES FOR PRODUCING 100 X 10^6 SCF/day HYDROGEN (90% Stream Factor; Annual Costs)

Process	Steam Reforming	Partial Oxidation	Koppers-Totzek	Steam-Iron	Electrolysis Current	Electrolysis SPE
Raw Material	Natural Gas	Residual Oil	Wyoming Coal	Montana Coal	Power	Power
Raw Material Unit Cost	$2.00/$10^6$ Btu	$15/bbl	$20/ton	$17.70/ton	3c/kWhr	3c/kWhr
Raw Material Unit Cost, $/$10^6$ Btu	$2.00/$10^6$ Btu	2.35	1.17	1.00	8.79	8.79
Overall Thermal Efficiency, %	74	82.7	59.4	62.6[b]	75.7	77.6
Facilities Investment, $$10^6$	51	146.2	249.4	197	270.6	74.3
Capital Required, $$10^6$	66	178.3	298.8	237	342.0	109.9
Gross Revenue, $$10^6$	56.38	88.75	120.79	109.39	232.22	153.85
By-Product Credits, $$10^6$	3.3[a]	(0.10)	(0.06)	24.96[c]	(6.96)[d]	(6.96)[d]
Net Revenue, $$10^6$	53.08	88.65	120.73	85.43	225.26	146.89
H$_2$ Production, 10^6 SCF/year			33,000			
H$_2$ Price, $/$10^3$ SCF	1.61	2.69	3.66	2.56	6.83	4.45
H$_2$ Price, $/$10^6$ Btu	4.76	8.23	11.20	7.84	20.90	13.62

a Steam at $3.00/1000 lb.
b Hydrogen plus heat equivalent of electric power.
c Electricity at 2c/kWhr.
d Oxygen at $10/ton.

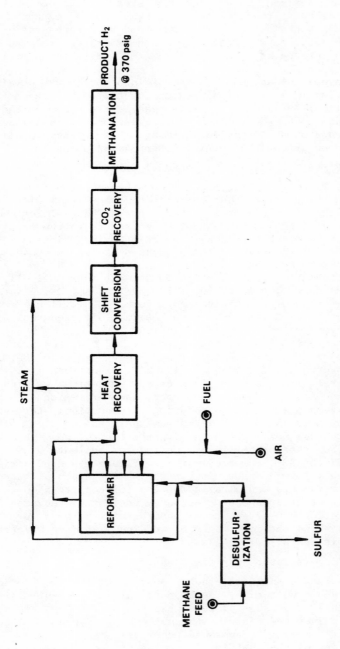

Figure 1. Hydrogen from conventional steam–methane reforming—simplified flow diagram

monoxide in the effluent is reacted with steam by the well-known
carbon monoxide shift reaction to yield hydrogen and carbon
dioxide; the carbon dioxide is then scrubbed by a regenerative
liquid scrubbing process. Residual carbon oxides are removed by
conversion to methane in a catalytic reactor. Hydrogen purity is
about 98%.

Basic investment and operating cost information was obtained
from information in our files plus some input from Reference (1)
adjusted to mid-1979 dollars. Natural gas is the assumed feedstock at a base cost of $2.00/$10^6$ Btu. Based on the guidelines,
the capital cost of a 100 million SCF/day plant is $66.3 million,
and the hydrogen price is $1.61/$10^3$ CF or $4.76/$10^6$ Btu.

Hydrogen By Partial Oxidation Of Residual Oil

Figure 2 is a simplified block flow diagram for producing
hydrogen from the partial oxidation of the residual oil with steam
and oxygen using the Shell or Texaco Process. The gasifier
effluent, at about 500 psig and 2300°F, goes through the steps of
waste heat recovery, soot removal, carbon monoxide shift, acid
gas removal, and methanation. The product hydrogen is at 400 psig
and contains 97.5% hydrogen.

Basic information on facilities investment, residual oil,
utilities, and operating labor for a 100 million SCF/day plant was
taken from Reference (1). Capital cost for a 100 million SCF/day
hydrogen plant is $178 million. With residual oil feedstock at
$15/barrel, hydrogen price is $2.69/$10^3$ SCF.

A recent presentation by the developers of the same process
(3) gives the capital investment for a hydrogen plant at $85 mil-
lion, and a hydrogen price of $2.22/$10^3$ SCF, significantly lower
than the figure assumed above. On closer examination, we find
that this is an installed equipment cost, however, which does not
include an oxygen plant, interest during construction, start-up,
or working capital. Adding these raises the total to $110 million.
Oxygen is purchased for $32/ton. Using the capital and operating
cost inputs from their paper in our method, we evaluate a hydrogen
price of $2.57/$10^3$ SCF, compared with the figure of $2.69 derived
from Reference (1). Incorporating an estimated oxygen plant cost
plus our other factors would raise the capital cost to about
$190 million. If steam and cooling water facilities required for
the oxygen section are added, total capital will be near our
number. This shows the importance of working with data which are
on a comparable basis.

Hydrogen By Koppers-Totzek Gasification Of Coal

Figure 3 is a simplified block flow diagram for the production of hydrogen from coal by the Koppers-Totzek Process. This
is a commercially available process, based upon rapid partial
oxidation of pulverized coal in suspension with oxygen and steam
at essentially atmospheric pressure under slagging conditions.
The raw gas leaving the gasifier is cooled to recover waste heat,

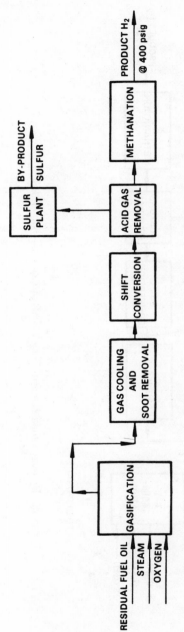

Figure 2. *Hydrogen from residual fuel oil partial oxidation—simplified flow diagram*

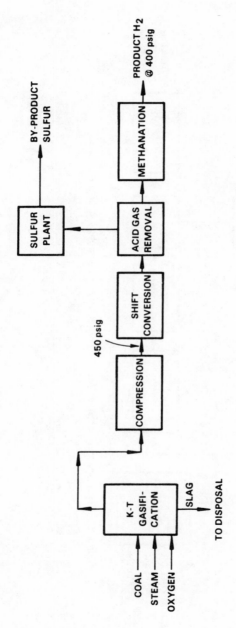

Figure 3. Hydrogen from Koppers–Totzek coal gasification—simplified flow diagram

then quenched with water to remove entrained ash particles before going through the steps of compression, carbon monoxide shift, acid-gas removal, and methanation. For this case, the product hydrogen is at about 400 psig and contains 97.5% hydrogen.

Basic information on facilities investment, coal, utilities, and operating labor for the 100 million SCF/day base case was obtained from Reference (1) and adjusted to a mid-1979 basis. Based upon the economic evaluation guidelines for 100 million SCF/day hydrogen production with Wyoming coal at \$20/ton (\$1.17/10^6 Btu), the total capital required is \$298.8 X 10^6 and the hydrogen price is \$3.66/10^3 SCF, or \$11.20/10^6 Btu. For Illinois coal at \$20/ton, total capital required is \$270.1 X 10^6 and hydrogen price is \$3.19/10^3 SCF, or \$9.76/10^6 Btu.

Hydrogen By The Steam-Iron Process

The production of hydrogen by the Steam-Iron Process is shown schematically in Figure 4. Although a coal-based process, hydrogen by the Steam-Iron Process is derived by the decomposition of steam by reaction with iron oxide, rather than synthesis gas generated from coal. Coal is gasified to provide a producer gas for the regeneration of iron oxide. Because hydrogen is not derived or separated from the producer gas, air can be used in the gasifier; nitrogen cannot significantly contaminate the hydrogen because of the iron oxide "barrier."

The iron oxide circulates between oxidation and reduction zones. The following reactions are typical of those occurring in the steam-iron reactor section at temperatures in the 1500° to 1600°F range.

Reductor: $Fe_3O_4 + CO \rightarrow 3FeO + CO_2$
              $Fe_3O_4 + H_2 \rightarrow 3FeO + H_2O$

Oxidizer: $3FeO + H_2O \rightarrow Fe_3O_4 + H_2$

The oxidizer effluent contains 37% hydrogen and 61% steam plus small amounts of nitrogen and carbon oxides. Condensation of the steam leaves a gas containing 95.9% hydrogen, 1.6% carbon oxides, and 2.5% nitrogen. No carbon monoxide shift or acid-gas scrubbing are needed. A cleanup methanation step reduces carbon oxides to 0.2% followed by drying and compressing to 325 psig to give product gas.

The carbon monoxide and hydrogen in the producer gas are not completely converted in reducing iron oxide. Heating value plus sensible heat at 1520°F in the gas exiting from the reductor represent 54% of the input coal fuel value. Part of this energy, 15% of coal fuel value, is utilized in the plant to compress air and generate steam. A larger amount is used to generate electric power. After dust removal with cyclone separators and electrostatic precipitators, the effluent gas, at 1520°F and 365 psia, is used to fuel a gas turbine system. The gas is burned in a combustor with excess air at 290 psia. (A 75 psi control valve loss is assumed.) The effluent from the combustor, at 2400°F, is expanded to 16.5 psia and 1198°F in the gas turbine. A portion

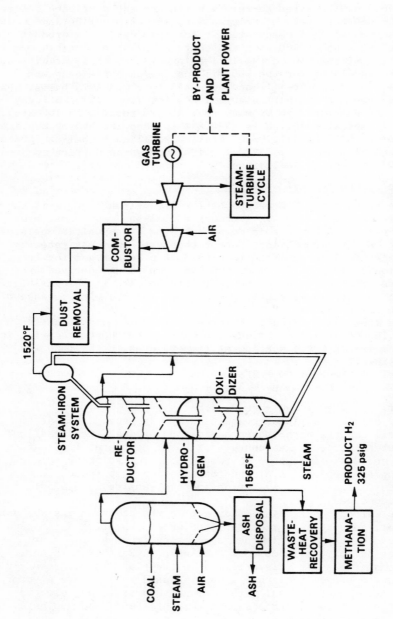

Figure 4. Hydrogen from steam–iron process—simplified flow diagram

of the expansion power is used to drive the combustor and producer
air compressors. The remaining shaft power is converted to electric power in a generator. Currently, the maximum temperature
range is 1800° to 2000°F, but gas turbines with an inlet temperature of 2400°F are expected to be available by the time this
process becomes commercial.

The expanded gas is used in a steam-power cycle to generate
electric power from 2400 psig, 1000°F steam generated in a waste-
heat boiler, and to generate a portion of the required process
steam for the oxidizer. The cooled combustor gas leaves the steam
cycle waste-heat boiler at 350°F. A total of 1325.4 MW of power
is generated in addition to the shaft power used for air compression; approximately 7% to 8% of this energy is used in the plant
for motor drives, etc. With this design adapted to an output of
100 million SCF/day of hydrogen, the by-product electric power
amounts to 158 MW. The base credit value assumed for this is
2¢/kWhr.

Basic investment and operating cost information were scaled
down to 100 million SCF/day from a larger IGT design (2). The
total capital cost, according to the guidelines set in Table 1,
is $237 million. At the base coal cost of $17.7/ton ($1.00/10^6
Btu), hydrogen price is $2.56/$10^3$ CF, or $7.84/$10^6$ Btu.

Hydrogen By Water Electrolysis

In water electrolysis, alternating current is rectified to
direct current and fed to a battery of electrolyzer cells.
Hydrogen and oxygen are released separately and the gases are
manifolded from the cell batteries into separators, where water
vapor and excess electrolyte are removed. The system costs
included feedwater treatment and hydrogen compression to 400 psi.
A block flow diagram is shown in Figure 5.

Two cases were examined for the production of water electrolysis. Data were taken from Reference (1) and adjusted to mid-1979
levels in accordance with Table 1. The costs of "current technology" electrolysis were averaged in Reference (1) from information provided by Lurgi, Electrolyser Corp., General Electric, and
Teledyne Isotopes. An advanced electrolyzer design, based upon
the General Electric Solid Polymer Electrolyte (SPE) design, was
also addressed as the second case.

In the source reference, Chem Systems have assumed an overall
efficiency, power-to-hydrogen, of 75.7% and 77.6%, respectively,
for current and SPE technology. The goals of the General Electric
SPE program are to achieve a 90% efficient system — already
demonstrated on the bench scale — and to reflect this, we have
taken a third case in which the hydrogen costs were recalculated
using a 90% efficiency. Assuming that such an efficiency might
be achieved commercially, both capital and operating costs are
lowered: The facilities investment and electric power requirements were reduced by a factor of 77.6/90 = 0.862 from the original
design.

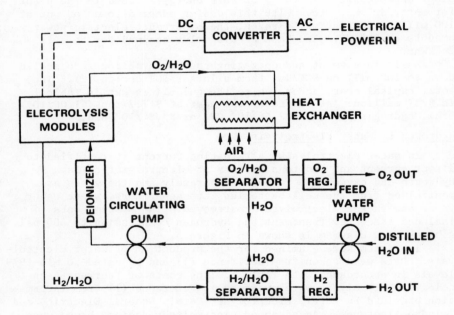

Figure 5.   SPE electrolytic hydrogen generating plant—simplified schematic

Thermochemical Hydrogen Production

If hydrogen is to be used as a carrier of energy produced from nuclear or solar furnaces, there is an incentive to develop processes for the thermal dissociation of water. The problem of splitting water using thermal energy alone within the temperature constraints of practical nuclear and solar heat sources has been addressed by the concept of thermochemical water splitting. In this concept, a multi-stage cyclic chemical process would be operated in which water and heat are the only inputs and hydrogen, oxygen, and degraded heat are the only outputs. To date, this technology is receiving wide attention, but it is still in its infancy and is thus difficult to make sound economic projections. Literally hundreds of cycles have been proposed and analyzed mainly on the grounds of efficiency and operability. Some of the more promising cycles involve both thermal and electrochemical steps and are known as "hybrid" cycles. In these, some of the input heat is used to produce electric power which in turn is used to drive an electrochemical reaction. For a few promising cycles, stepwise, bench-scale demonstrations of the chemistry have been performed, and some flowsheet analysis has been carried out. For the purposes of this paper, one cycle has been selected on the basis of having had more economic study performed on it than most of the others. The cycle is known variously as the "Westinghouse Sulfur Cycle" and the "EURATOM Mark II-V6 Cycle." It is a thermal-electrochemical hybrid cycle as follows:

$$2H_2O + SO_2 \rightarrow H_2SO_4 + H_2 \quad \text{(electrolysis)}$$
$$H_2SO_4 \rightarrow H_2O + SO_3 \quad \text{(thermal)}$$
$$SO_3 \rightarrow SO_2 + \tfrac{1}{2}O_2 \quad \text{(thermal)}$$

The process is assumed to operate on thermal energy delivered as high-temperature helium from a high-temperature, gas-cooled nuclear reactor (VHTR) operating at about 1742°F.

An economic analysis was carried out by Knoche and Funk (4) based upon a flowsheet developed by Westinghouse. This analysis indicated a projected hydrogen cost of about $6.50/$10^6$ Btu based upon the following fundamental assumptions:

- Cost of nuclear heat: $1.50/$10^6$ Btu
- Process efficiency: 45%
- Capital cost of non-nuclear portion of the plant: $425 X 10^6
- Plant size: 360 X 10^6 SCF/day hydrogen
- Nuclear reactor size: 3345 MW (thermal).

Because of the considerable uncertainties in all the assumptions made, this projected hydrogen cost should not be rigorously compared with the cost projections we have made for the other processes discussed in this paper. Moreover, we have not reworked Knoche and Funk's economics to conform with the guidelines of Table 1. (The assumptions they made were an 80% stream factor, utility financing, $12\tfrac{1}{2}$% capital recovery factor, and mid-1976 dollars.)

A completely independent assessment of the same cycle was made by Broggi et al. (5) Their estimated hydrogen cost was about $9/10^6$ Btu based upon process heat supplied at $2.50/10^6$ Btu. The selected cost of nuclear process heat appears to be somewhat arbitrary, although in Knoche and Funk's case, they derived the cost from a composite of fuel, operating, maintenance, and invest- ment components.

If solar heat is considered for a thermochemical hydrogen production process, two additional variables must be considered. The temperature limitations on a solar furnace are controlled by factors other than those for a nuclear reactor, and it is possible to consider temperatures in excess of 1742°F, normally considered to be the limit of HTGR technology. However, it is important to know how the cost of heat varies as a function of the delivery temperature, and no data for this variation are available for solar furnaces in the 1292° to 2732°F range. Second, because a solar furnace will only operate intermittently, the plant factor will be considerably less than 100%. No analysis has been published to date for the sensitivity of cost of a thermochemical hydrogen process to plant factors in the 20% to 40% range. Al- though the production of hydrogen from solar thermal heat is an interesting and potentially important technology, it has not yet progressed to the stage where even educated guesses can be made as to its cost.

Comparison Of Processes

Table III presents a detailed summary of investment and annual operating costs for the different processes. These are "base cases" from which the sensitivity analysis of various factors has been derived. These can be considered as representative of current costs for the purpose of process comparison.

Sensitivity Analysis. The sensitivity of hydrogen price to the following variables was calculated: raw materials cost, plant capital cost, stream factor, plant size, and by-product credits. The base case assumptions are shown in Table III. The methodology used for these sensitivity calculations was as follows:

• Raw Materials Cost: We assumed two or more raw material costs and then calculated the hydrogen price using the following procedure for each raw material cost:

Raw Material	Process	Raw Material Price		
		Below	Base	Above
Residual Fuel Oil	Partial Oxidation	$10/bbl	$15/bbl($12.35/10^6$ Btu)	$20/bbl
Wyoming Coal	Koppers-Totzek Process	$10/ton	$20/ton($1.17/10^6$ Btu)	$30/ton
Electricity	Electrolysis	1¢/kWhr	3¢/kWhr($8.79/10^6$ Btu)	5¢/kWhr
Montana Coal	Steam Iron	--	$17.7/ton($1.00/10^6$ Btu)	$1.50/10^6$ Btu
Natural Gas	Steam Reforming	--	$2.00/10^6$ Btu	$4.00/10^6$ Btu

Table III. SUMMARY OF CAPITAL AND OPERATING COSTS FOR PRODUCING 100 X 10^6 SCF/DAY OF HYDROGEN AT 300 TO 500 psig BY VARIOUS PROCESSES (Mid-1979 COST BASIS, 90% Stream Factor)

Process	Steam Reforming (Natural Gas at $2.00/10^6 Btu)	Partial Oxidation (Residual Oil at $15/bbl)	Koppers-Totzek Gasification (Wyoming Coal at $1.17/10^6 Btu)	Steam-Iron (Montana Coal at $1.00/10^6 Btu)	Electrolysis (Current) (Electric Power)	Electrolysis (New Technology) (Electric Power)
Annual H₂ Production, 10³ SCF				33,000		
Capital Investment ($/10⁶)						
Facilities Investment	51.0	146.2	249.4	196.7	270.6	74.3
Interest During Construction	7.6	21.9	37.4	29.5	40.6	11.3
Start-Up Costs	1.0	2.9	5.0	3.9	5.4	1.5
Working Capital	6.7	7.3	7.0	7.2	25.4	22.8
Total Capital Required	66.3	178.3	298.8	237.3	342.0	109.9
Capital Required $/daily 10³ [a]	663	1783	2988	2373	3420	1099
Annual Operating Costs, $/10³ [a]						
Raw Materials	33,000	30,195	21,673	24,970 [b]	123,351	120,247
Utilities						
Power	1029	339	--	(Integral)	--	--
Process Water	156	482	2,765	100	1,143 [c]	695 [c]
Cooling Water	221	--	--	--	--	--
Stack-Gas Cleanup	--	373	--	--	--	--
Operating Labor	306	468	540	1,890	360	180
Operating Labor Supervision	46	70	81	284	54	27
Maintenance						
Labor	1020	2,924	4,988	3,934	5,412	1,486
Supervision	153	439	748	590	812	223
Materials	1020	2,924	4,988	3,934	5,412	1,486
Administrative and Support Labor	509	780	1,271	2,126	1,328	383
Payroll Extras	611	936	1,526	2,552	1,593	460
Operation and Maintenance for Combined Cycle By-Product Power Section	--	--	--	2,180	--	--
Insurance	1020	2,924	4,988	3,934	5,412	1,486
General Administrative Expenses	1020	2,924	4,988	3,934	5,412	1,486
Depreciation	2983	7,310	12,470	11,505	13,530	3,715
Return on Investment	13,264	35,660	59,760	47,460	68,400	21,980
Gross Revenue Required	56,359	88,748	120,786	109,393	232,219	153,854
Less By-Product Credits						
Sulfur, $10/long ton	--	-100	-64	--	--	--
Steam, $3.00/1000 lb	-3300	--	--	--	--	--
Power, 2c/kWhr	--	--	--	-24,960	--	--
Oxygen, $10/ton	--	--	--	--	-6,960	-6,960
Net Revenue Required	53,058	88,648	120,722	84,433	225,259	146,894
H₂ Price, $/10³ SCF	1.61	2.69	3.66	2.56	6.83	4.45
H₂ Price, $/10⁶ Btu	4.76	8.23	11.20	7.84	20.90	13.62

a Unit costs from Table 1.
b Includes coal (24,031) plus catalysts, chemicals, turbine blades, and vanes.
c Derived as total utilities from publication.

- Investment Changes: We assumed two <u>facilities investment costs</u>, one 20% above and the other 20% below the base case, and calculated the hydrogen price using the same procedure as for the base case.
- Stream Factor Changes: We assumed that for stream factors of 30%, 50%, and 70%, the raw materials requirement would be reduced proportionately, but capital-related costs do not change from the base case. We calculated the hydrogen price the same way as in the base case.
- Capacity Changes: Plant capacities of 25 and 50 million SCF/day were considered. The power factors on capacity were assumed as follows:

Process	Size, mm	Power Factor
Koppers-Totzek, Partial Oxidation	100-50	0.7
Koppers-Totzek, Partial Oxidation	50-25	0.6
Electrolysis	100-25	0.9
Steam Iron	100-25	0.65
Steam Reforming	100-25	0.65

  The raw materials, utilities, and by-products were changed directly by the ratio of plant capacity. The hydrogen price was calculated using the same procedure as in the base case. Variations between the 0.6 and 0.7 power factor have only a small effect on hydrogen price. The assumption of many modules for electrolysis suggests a higher exponent.
- By-Product Credit Changes: Only two processes — steam-iron and electrolysis — are significantly affected by by-product credits. Steam-iron has electric power as a by-product, which was calculated at 2¢ to 4¢/kWhr, and electrolysis has oxygen as a by-product, which was varied from $10 to $30/ton.

Presentation Of Results

Because the data for the thermochemical process are somewhat speculative and are not considered sufficiently sound to warrant conversion to the guidelines spelled out in Table 1, data for this process are exhibited separately in Figure 12. Data for all the other processes have been aggregated in Figures 6 through 11.

Figure 6 shows the sensitivity of hydrogen price to feedstock price, all plotted on the same scale. It should be noted that the "feedstock" for electrolysis is electric power, not primary fuel, so that capital costs and inefficiencies associated with power generation are included in the raw material cost. The circles on this figure represent the base case for our calculations and reflect approximately realistic values for raw material costs for large-scale plants in 1979.

Figure 7 shows the sensitivity of hydrogen price to variations in plant cost, and also exhibits the considerable differences in plant cost assumed for the base case — believed to be reasonable

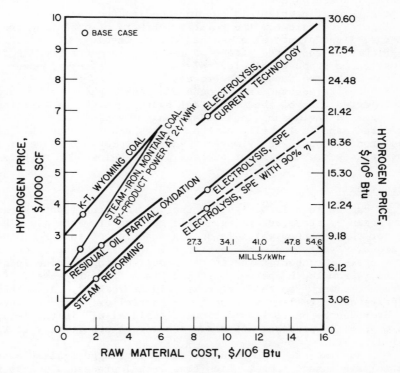

Figure 6. Effect of raw materials cost on product price for 100×10^6 scfd by various processes

estimates of 1979 values. Again, the electrolysis plant costs do not include costs associated with electric power generation.

Figure 8 is based upon essentially the same data as Figure 7 but plotted as a sensitivity to percentage change in estimated facilities cost. This emphasizes the greater sensitivity of the more capital-intensive processes to this parameter.

Figure 9 shows how the cost of hydrogen may be expected to change as the scale of the plant is reduced from the 100×10^6 SCF/day base case. Some subjective judgments were made as to the relative scale factors for electrolysis and the other processes. Electrolysis scales down more favorably primarily because the electric generating plant is not included in the system.

Figure 10 indicates the relative effects of operating the base-case-sized plants at a reduced output level.

Figure 11 shows the effect of taking by-product credits for the oxygen co-produced by electrolysis and the electric power co-produced by the Steam-Iron Process.

Figure 12 (adapted from Knoche and Funk's paper) shows how the projected cost of thermochemical hydrogen might be affected by changes in the assumptions made. Only one plant size is considered, and it is believed that plants smaller than this size would produce considerably more expensive hydrogen because of the unfavorable scale factors of nuclear plants below 3000 MW_{th} in size.

Conclusions

It can be seen from Figure 6 that by far, the least expensive is hydrogen by steam reforming. This results from the low plant cost for this system, which is only a fraction of that for other processes. Even though the cost of the natural-gas feed, on an energy basis, is about twice that of coal, hydrogen price is about half that from the Koppers-Totzek and Steam-Iron Processes, if by-product power from the latter is sold at 2¢/kWhr. At 4¢/kWhr, the price drops to less than that for partial oxidation (Figure 11).

Hydrogen by partial oxidation of residual oil is the second most economical process. It shows the lowest sensitivity to raw material cost.

We can also see from Figure 6 that, for large-scale plants, electrolysis cannot compete with fossil-fuel processes unless the power costs are far below today's base-load industrial rates. We can also see the significant reductions in hydrogen costs that would be achieved if the goals of the SPE electrolyzer program are met; but even so, electrolysis could not compete with natural gas reforming unless gas prices rose to about $6.00/10^6$ Btu. It appears, however, that electrolysis from power sources priced at the 20 to 30 mills/kWhr range can hope, with development, to compete with coal-based hydrogen processes if coal costs are $2.00/10^6$ Btu or more. Under today's conditions of price, hydrogen from gas and oil has an almost unbeatable cost advantage over any other source.

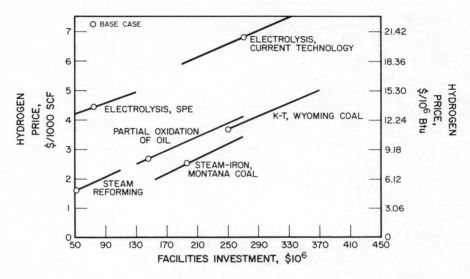

Figure 7.   *Effect of facilities investment on hydrogen price for 100 × 10⁶ scfd of hydrogen by various processes*

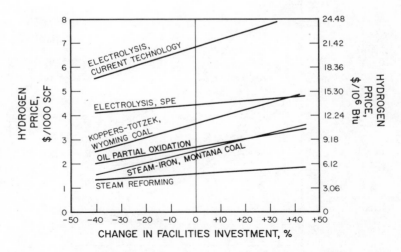

Figure 8.   *Effect of change in facilities investment on hydrogen price for 100 × 10⁶ scfd of hydrogen production by various processes*

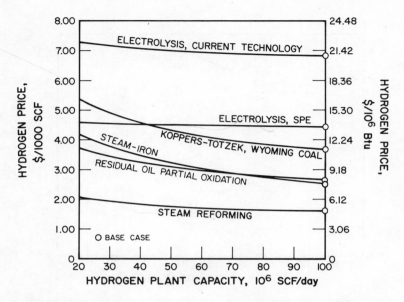

Figure 9.   Effect of plant capacity on hydrogen price by various processes

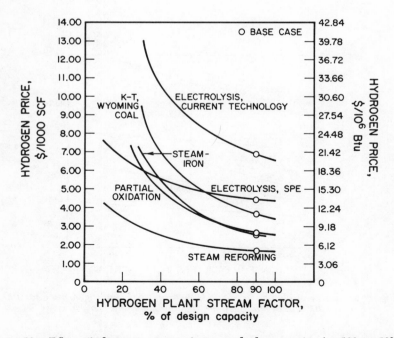

Figure 10.   Effect of change in stream factor on hydrogen price for 100 × 10⁶
scf/day of hydrogen production by various processes

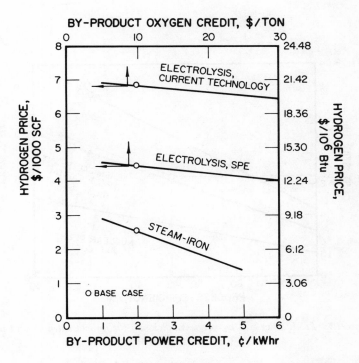

Figure 11. Effect of by-product credits on hydrogen price for 100 × 10⁶ scfd of hydrogen by electrolysis and steam–iron processes

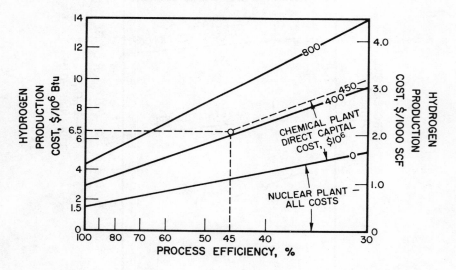

Figure 12. Production cost vs. process efficiency for hybrid sulfuric acid process
(360 × 10⁶ scfd)

The future attractiveness of the coal-based process will depend upon the availability of natural gas or oil at a reasonable cost. If the coal cost rises to $1.50/10^6 Btu from the $1.00 level, then the cost of natural gas must rise to $3.10/10^6 Btu for hydrogen by the Steam-Iron Process to compete with reforming if by-product power sells at 4¢/kWhr. At 2¢/kWhr by-product power, natural gas cost must rise to $4.65/10^6 Btu for the Steam-Iron Process to be competitive.

Figure 7 shows that if SPE electrolysis meets its development goals, minimal capital investment would be required to install an electrolyzer and rely upon purchased electric power. Figures 7 and 8 both suggest that considerable advantages would be obtained if capital costs of the coal-based processes could be reduced.

Figure 9 relies upon some arbitrary assumptions of scale factor, but indicates why small hydrogen plants would favor oil, gas, and electrolytic processes rather than coal-based systems. A more important conclusion is reached from Figure 10, which indicates that there would be severe penalties in operating the coal and oil plants intermittently. It also indicates that even electrolysis, which has been promoted as an off-peak power user, would be very expensive if operated below the 20% to 30% plant factor that is associated with "cheap" off-peak power. The huge reduction in capital cost promised by the SPE electrolyzer offsets this somewhat, but operation of even an advanced electrolyzer at less than 20% of the time (5 hours/day) is unattractive compared with a dedicated, full-time reformer or gasifier. The low capital cost of steam reforming is again shown to advantage.

Figure 11 indicates the significant effect of properly allocating by-product power credits to the Steam-Iron Process, and less sensitive, though still significant, is the oxygen credit for electrolysis. It is interesting that hydrogen produced by Steam-Iron, taking a 2¢/kWhr credit for power, costs about the same (at about $2.60/10^3 CF) than hydrogen produced from an advanced SPE electrolyzer operating on this 2¢ power (about $2.75/10^3 CF from Figure 6), so that a combination Steam-Iron/electrolyzer plant appears to have merit.

The cost of hydrogen derived in Figure 12 from thermochemical processes is in the same order as the current cost of hydrogen produced from partial oxidation of oil. Even if the cost of nuclear process heat is three times the assumed value, the cost of hydrogen is still only about $11/10^6 Btu, cheaper than advanced SPE electrolysis. The accuracy and credibility of the data shown in Figure 12 are questionable because the process itself is still at the exploratory level at the lab bench. However, the process merits careful investigation because of this observation.

Abstract

There is no simple answer to the question "How much does
hydrogen cost to produce?" Hydrogen production today takes place
from natural gas, oil, and coal as feedstocks, and by electrolysis,
each process having conflicting and contrasting economics. In the
future, hydrogen production from nuclear energy and the various
forms of solar energy are contemplated, and new improved technologies for all of these production routes are under development.

This paper discussed the relative sensitivities of hydrogen
cost to capital, feedstock, and utility prices for various different types of production processes, outlines some process
selection criteria for different applications, and indicates what
changes in process economics might be anticipated from new technical developments.

Literature Cited

1.   Corneil, H.G., Heinzelmann, F.J. and Nicholson, W.S.,
     "Production Economics for Hydrogen Ammonia and Methanol
     During the 1980-2000 Period," Exxon Research & Engineering
     Company Report No. BNL-50663, Linden, New Jersey, April
     1977.

2.   Tsaros, C.L., Arora, J.L. and Burnham, K.B., "A Study of the
     Conversion of Coal to Hydrogen, Methane, and Liquid Fuels
     for Aircraft," prepared for NASA under Contract No. NAS1-
     13620, June 1976.

3.   Reed, C.L. and Kuhre, C.J., "Hydrogen Production From Partial
     Oxidation of Residual Fuel Oil." Paper presented at the
     Symposium on Production and Marketing of Hydrogen, Current
     and Future, ACS/CSJ Chemical Congress, Honolulu, Hawaii,
     April 1-6, 1979.

4.   Knoche, K.F. and Funk, J.E., "Entropy Production, Efficiency,
     and Economics in the Thermochemical Generation of Synthetic
     Fuels. I. The Hybrid Sulfuric Acid Process," Int. J. Hydrogen
     Energy 2, No. 4, 377-386 (1977).

5.   Broggi, A., Joels, R., Pertel, G. and Morbello, M., "A
     Method for the Technoeconomic Evaluation of Chemical
     Processes — Improvements to the 'Optimo' Code," Proceedings
     of the First Technical Workshop on Thermochemical Processes,
     Document H-1-1, Paper I-223, International Energy Agency,
     Ispra, Italy, August 28-31, 1978.

RECEIVED July 12, 1979.

DOE Program on Hydrogen Energy Systems

BEVERLY J. BERGER and JAMES H. SWISHER

Division of Energy Storage Systems, U.S. Department of Energy, Washington, D.C. 20585

The U.S. Department of Energy has a broad-based effort on hydrogen energy systems. While it is not a main thrust of the DOE energy program at present, the level of effort on hydrogen technology put forth by DOE is expected to increase as work is completed on programs with more potential for near-term oil savings. The scope of our activities is depicted in Figure 1, where the activities are subdivided into production, storage, transport, and conversion technologies. It is generally recognized that the production of hydrogen requires more effort than the other areas of hydrogen technology at present for two reasons. First, storage, transport, and conversion technologies will not be used if economic processes for hydrogen production are not developed. Second, some of the most difficult technological problems are associated with hydrogen production.

Note that energy sources and energy carriers available for hydrogen production include coal, electricity, nuclear process heat, and solar energy. Storage options include hydrides, liquid hydrogen, gasous hydrogen in pressurized tanks, and storage in underground caverns. Hydrogen may be converted to electricity in fuel cells and turbines or used directly as a chemical feedstock and multipurpose fuel. Use of existing gas pipelines will be important for hydrogen use. A strong program is developing to establish pipeline compatibility with hydrogen. Another key technological advance is the development of lightweight, low-cost hydrogen storage for vehicles. Such storage is required before hydrogen can have wide spread use as a non-polluting vehicle fuel.

DOE Hydrogen Budget

In fiscal 1979, DOE allocated approximately $43 million for hydrogen R&D (Table I). In the coal conversion area the planned budget was $22.8 million, which is almost half of the $43 million; the largest part of that $22.8 million was for building a pilot plant for producing hydrogen from coal. However, the plant was not built because of difficulties in contract negotiations and

This chapter not subject to U.S. copyright.
Published 1980 American Chemical Society

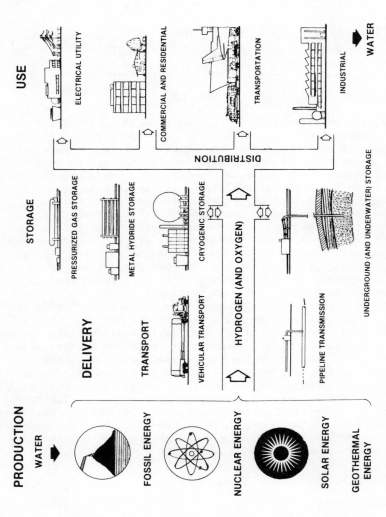

Figure 1.   Hydrogen energy system

Table I. Budget Data for DOE Hydrogen Program

	FY 78 (Planned) $M	FY 78 (Actual) $M	FY 79 $M	Presidential Budget Request $M
CC	22.8	2.2	0	0
BES	8.4	same	8.9	9.6
MA	2.3	"	2.5	2.7
STOR	6.3	"	6.3	2.5
LF	1.0	"	1.0	1.0
MFE	0.6	"	0.9	1.2
Solar	1.2	"	1.1	1.1
TEC	0.2	"	0.2	0.2
EV	0.3	"	0.3	0.3
RA	0	"	0.5	0
TOTAL	43.1	22.5	21.7	18.6

Key:
CC = Coal Conversion
BES = Basic Energy Sciences
MA = Military Applications
STOR = Energy Storage Systems
LF = Laser Fusion
MFE = Magnetic Fusion Energy
Solar = Solar Technologies
TP = Transportation Program
EV = Environmental Sciences
RA = Resource Applications

funding priorities. There is no funding available or planned for
hydrogen from coal in FY 1979 of FY 1980.

Basic Research on Hydrogen

The Division of Basic Energy Sciences has supported a hydrogen program for several years. The budget for this year is $8.9
million. Emphasis is placed on basic research that parallels the
development and demonstration projects being carried out by other
DOE divisions. The goal is to understand phenomena rather than
invent devices. A substantial amount of research is sponsored on
producing hydrogen thermochemically and by other solar-assisted
processes. Research to improve catalysts and to improve electrocatalyst life and performance are supported, not only because of
the effect such improvements could have on hydrogen production,
but also because of their importance for battery and fuel cell
development. The Division of Basic Energy Sciences has also
funded work on hydrogen embrittlement phenomena.

Applied Research on Hydrogen

The broadest hydrogen program lies in the Division of Energy
Storage Systems and includes production of hydrogen from water,
storage of hydrogen, transmission and distribution, and some
aspects of end use. This program was funded at $6.3 million in
FY 1978 with an increase to $7.4 million planned for FY 1979.
However, $6.3 million was provided in FY 1979 and the Presidential
budget request to Congress for FY 1980 is $2.5 million. Within
the FY 1980 budget request, only those projects which are covered
by the existing International Energy Agency Hydrogen Agreement
will continue to be funded. These include production of hydrogen
from water by electrolysis and thermochemical cycles and a market
study.
There is a small hydrogen effort within the Laser Fusion and
Magnetic Fusion programs focused on the integration of hydrogen
production within fusion systems. Solar divisions now fund
advanced research projects for the direct production of hydrogen
from sunlight and solar-assisted thermochemical hydrogen production. DOE currently has no major projects to produce hydrogen
from solar energy.* A few projects on the use of photovoltaic
devices and biomass for hydrogen production are being managed by
the Solar Energy Research Institute (SERI) for DOE. The possibility of using ocean thermal energy conversion (OTEC) for hydrogen production is also under study.
The transportation program supports research on hydrogen-
fueled vehicles and has complemented the efforts of the Division
of Energy Storage Systems in developing hydrogen storage systems
for vehicles. Under the Assistant Secretary for Environment
about $300,000 is expended each year to evaluate health, environ-
ment, and safety issues related to hydrogen production and use.

In FY 1979, $0.5 million was expended for a low-head hydro
facility to produce hydrogen electrolytically by the Office of
Resource Applications under joint sponsorship with the Division
of Energy Storage Systems.

   Internal coordination of the DOE hydrogen program is accomplished through the DOE Hydrogen Energy Coordinating Committee.
The members of the committee are key individuals from each
division which has programs involving hydrogen technology, either
directly as in the Division of Energy Storage Systems or
indirectly as in Fusion Energy Systems. The committee meets once
every six weeks to exchange ideas, R&D results and future plans.
DOE officials are kept informed of interests and developments in
the hydrogen community. It is understood that hydrogen may play
a key role in the U.S. energy future; the time frame when hydrogen is likely to make a significant impact is not at all clear.
This will be determined largely by the continuing results of the
research and development which is going on today. The DOE hydrogen program has an opportunity to affect the nation's future by
supplying a clean burning fuel for use in homes, industries and
vehicles. In the near to mid term the likely impact will be in
the chemical market place, reducing the amount of natural gas
required to make hydrogen for use as a chemical. Use of hydrogen
as a fuel will come later. Let us hope that our research and
development is successful.

RECEIVED July 12, 1979.

Hydrogen Technology: An Overview

F. J. SALZANO, A. MEZZINA, M. BELLER, G. STRICKLAND, and S. SRINIVASAN

Brookhaven National Laboratory, Upton, NY 11973

Hydrogen's potential to be derived from renewable as well as from fossil resources such as coal, has caught the imagination of many, here and abroad. It is generally recognized that hydrogen's entry into the energy infrastructure, on a large scale, is not a near-term option. Yet, there exist opportunities for establishing closer relationships with complementary technologies such as solar and fuel cell development that may improve these prospects. Before hydrogen can begin to play the role of a universal fuel, developments in pertinent technology will be required; but these do not appear to be radical departures from the current state-of-the-art. This paper provides an overview of hydrogen technology and identifies the needs for further development, in order that hydrogen achieve its ultimate potential.

It should be noted that prior to 1950, substantial quantities of hydrogen mixed with CO were produced from coal and distributed in cities as town gas. Some cities in the world still distribute a 50% by volume mixture of H_2 with CO in city distribution systems. Currently, hydrogen is an industrial commodity, derived primarily from natural gas and the technology for handling it is familiar to industry.

Transmission

Hydrogen is presently shipped as a cryogenic fluid in insulated tank cars or trucks, or as a gas in tube trailers (manifolded banks of large compressed gas cylinders). Although hydrogen gas is not generally transmitted through long distance pipelines at present, the experience with natural-gas pipeline transmission permits extrapolation to hydrogen applications. Three possibilities are envisioned: (1) the existing natural gas pipeline system may be used for hydrogen; (2) a new pipeline system can be employed for hydrogen transmission; (3) hydrogen can be injected

0-8412-0522-1/80/47-116-033$05.00
© 1980 American Chemical Society

to form mixtures with natural gas in existing pipelines. In
the first case, the integrity of natural gas pipeline must be
established for handling hydrogen. For example, the metallurgical compatibility of hydrogen with pipe materials needs to be
established. Leakage due to its lower molecular weight and additional pumping costs needs to be considered. Secondly, new
pipeline systems present substantial capital costs. Finally,
compatibility of hydrogen-natural gas mixtures with gas appli-
ances as well as with the transmission/distribution network
must be examined.

The existing natural-gas system in the U.S. is complex; it
contains a total of about a million miles of pipeline, of which
two-thirds are used for distribution, one-quarter for transmission and storage, and the remainder for gas field and gathering
use. A wide range of transmission costs exists within this
maze, ranging from 1¢ to 95¢/10^6 Btu/100 miles. Table 1 and
Table 2 summarize a comparison of energy transmission costs and
distribution, respectively, for hydrogen vs. natural gas.

Large Bulk Storage

Natural gas transmission systems operate most efficiently
and at lower cost when run continuously at maximum capacity.
Therefore, at periods of low demand, gas is pumped to storage
reservoirs, preferably located in the consuming regions.
Later, the gas is withdrawn to serve peak loads, thus supplementing the carrying capacity of the main transmission system.
In 1975, 196 (79%) of the 376 underground storage reservoirs
were depleted gas wells. Alternative storage methods include:
depleted oil wells, salt mine caverns, aquifers, and above-
ground storage in cryogenic tanks.

The total underground reservoir capacity in the U.S. is
estimated at 2.3 x 10^{15} Btu of working gas. Underground caverns can also be employed for hydrogen storage. Using the volumetric ratio of heating value of three to one for natural gas
to hydrogen, the working gas capacity of the storage facilities
is in the order of 0.8 x 10^{15} Btu.

The use of existing underground caverns for storage is expensive. There are large investments in installing pipe systems to and from the reservoirs, and in compression equipment.
The total cost of service for underground natural gas storage
(excluding the cost of the gas) is estimated at about 75¢/10^6
Btu, an appreciable increment to production and transmission
costs.

Small Scale Storage

It should be noted that a key problem in the utilization
of hydrogen in some applications is the need for compact,
lightweight, and safe storage. This is especially true in

Table 1.

Estimated Transportation Costs per 100 Miles to a Local
Substation, Including Capital, Operating, and Maintenance Costs

Transportation	Cost ($ per 10^6Btu per 100 miles)
Methane by pipeline	0.035
Hydrogen by pipeline	0.105
Electricity by high voltage transmission	0.21
Gasoline by tanker	0.10

Table 2.

Distribution Costs from the Local Substation to a Residential
Household Including Capital, Operating, and Maintenance Costs

Distribution	Cost ($ per 10^6Btu)
Methane by pipeline	0.60
Hydrogen by pipeline	0.66
Electricity by overhead wire	2.55
Gasoline by truck	0.70

proposed automotive applications of hydrogen. Currently available techniques of high pressure gas storage are inadequate and the approach of using the cryogenic liquid has associated safety problems, as well as being energy intensive.

During recent years, the use of metal hydrides for hydrogen storage has been under development. Metal hydrides offer the potential of low cost and safe hydrogen storage through their ability to react with hydrogen to form decomposable hydrides. Heat, that must be supplied in order to decompose the hydride and release the hydrogen, is generally available at the point of hydrogen consumption.

Typically, a metal hydride formed from a 50–50 atom ratio of titanium and iron, currently the lowest cost and most convenient to use, offers a working storage capacity of approximately 1.5 wt % hydrogen. A desirable feature of $TiFeH_x$ is that the hydriding-dehydriding reactions occur near ambient temperatures. The reaction rates are very rapid but are limited by the rate of heat transfer. During the course of these reactions the charging pressure ranges between 500 and 15 psi. TiFe materials do not lose hydrogen storage capacity even after thousands of cycles when high-purity hydrogen is used. In circumstances where impurities in the hydrogen cause poisoning, the hydride can be reactivated by heating and evacuation. Oxygen, CO, and H_2O are common poisons.

Magnesium hydrides can store several times as much hydrogen per unit weight as TiFe; but are of limited application because heating to about $300^\circ C$ ($572^\circ F$) is required for dehydriding. In working systems, it should be possible to achieve a hydrogen content of about 5.5 wt %. The alloy cost is about one-half that of TiFe-based alloys, but the thermal load is nearly three times as high.

The high hydrogen concentration of Mg-based hydrides is an advantage in automotive storage where weight is an important factor. The elevated temperature requirements suggest the use of dual-hydride systems which employ a TiFe-based hydride for cold starting, and the use of hydrogen from the Mg-based hydride heated by exhaust gases. Working systems of this type have been demonstrated by Daimler-Benz AG of West Germany in a passenger car and a bus.

A third class of hydrides, the AB_5 type ($LaNi_5H_6$), also function at ordinary temperatures and are easily activated, but are costly due to the presence of lanthanum. Investigations seeking lower cost substitutes for lanthanum have resulted in the development of two useful alloys. One is based on the substitution of cerium-free Mischmetal (M), an unrefined rare-earth alloy; the second is based on the substitution of calcium (Ca) and Mischmetal. These compositional changes have reduced the raw materials costs to about 30% of that for $LaNi_5$ alloy, but further cost reductions are required.

A comparatively new and novel method of storing hydrogen
at very high pressure in hollow glass microspheres (microballoons) has been proposed. The technique is based on the high
permeation rate of hydrogen through certain glasses at temperatures of approximately $200^{\circ}C$ ($392^{\circ}F$), at pressures of 400
atmospheres. Like Mg-based hydrides, this storage technique
would function best for a transportation application when coupled with a TiFe-based hydride. A practical working system of
glass microballoons has yet to be demonstrated; However, this
material has the potential to be charged with hydrogen from
impure and mixed gas streams, based on diffusion rate differences between hydrogen and other gases.

Hydrogen Energy Systems and the Fuel Cell

Fuel cells may be characterized as modular, pollution-
free, high-efficiency, energy conversion systems that offer
flexibility of application over a wide range from kilowatts to
multi-megawatts. Fuel cell systems may exhibit differences in
electrolyte (acid, alkaline, solid polymer, molten carbonate,
solid oxides) and associated electrode structures and catalysts. These systems are at various stages of development with
cost/performance characteristics that can be matched to given
stationary and mobile applications. In all cases, the reactants are hydrogen and oxidizing agents--usually oxygen supplied from air--and in certain cases, chlorine or bromine.
Hydrogen is the best fuel for use in all fuel cells. In
those instances where the fuel cell system consumes natural gas
or liquid hydrocarbons, these fuels are first converted (steam
reformed) to hydrogen. Thus, a common element influencing cost
and performance of all fuel cells is hydrogen and its sources.
The use of external reformers allows for the chemical conversion of virtually any hydrocarbon fuel; however, substantial
benefits would accrue if pure hydrogen were made available. As
a result, when one considers hydrogen feed to a fuel cell it is
necessary to examine the relevant hydrogen technologies of production (from fossil, nonfossil and renewable energy resources)
and storage as well as pipeline transmission and distribution.
Hydrogen technology development incentives are derived
from the options offered in converting coal and nonfossil fuel
resources (solar, wind, hydro, nuclear) to a flexible fuel form
with excellent energy carrying efficiency. Hydrogen production
by coal gasification, by electrolysis and by thermochemical
decomposition of water represent the near-term, mid-term and
long-term production options, respectively.
Fuel cells in the most advanced stages of development are
illustrated in Table 3.

Utility-Stationary Applications. Fuel cell and hydrogen
energy technologies can be complementary over the full spectrum

Table 3.
Improvements in H_2/Air Fuel Cells

| | H_2-Air Alkaline | | Acid, H_2-Air | | | SPE, H_2-O_2 | |
	1967	1977	1967[1]	1977[1]	1977[2]	1967	1977
Current Density (mA/Cm²)	100	200	300	300	400	15-42	700
Cell Voltage (V)	0.75	0.7	0.55	0.65	0.6	0.6-0.7	0.68
Power Density (mW/cm²)	75	140	165	195	240	10-25	476
Operating Temp. (°C)	65-75	80	160	190	25	25	82-100
Thermal Efficiency (%)	51	47	37	44	41	40	46
Catalyst Loading (mg/cm²)	1	0.2	20	0.76	4	65	8
Startup Time	3.5	5M			15M	3 hrs	3-5M
Life (hrs)	5000	NA		12,000	600	1000	40,000

References:
Alkaline Fuel Cells: Kordesch, K.V., J. Electrochem. *Soc.* 118, 815 (1971).
Acid Fuel Cells: [1]Phosphoric Acid Electrolyte - Fickett, A.P., Fuel Cell
Electrolysis: Where Have We Failed? Proc. of the Symp. on Electrode Materials and
Processes for Energy Conversion and Storage, J.D.E. McIntyre, S. Srinivasan and F.
G. Will (Eds.), Vol. 77-6, 546-558. Electrochemical Society, Princeton, New
Jersey, 1977.
[2]Trifluoromethanesulfonic Acid (TFMSA) Electrolyte: Janskiewicz, S., George, M.
and Baker, B.S. Aqueous Sulfonic Acid Electrolyte Fuel Cell. National Fuel Cell
Seminar Abstracts, June 21-23, 1977, Boston, Massachusetts, 103-106.
Solid Polymer Electrolyte (SPE) Fuel Cells: McElroy, J. Status of Solid Electro-
lyte Fuel Cell Technology. National Fuel Cell Seminar Abstracts, San Francisco,
California, July 11-13, 1978, pp. 176-179.

of stationary and mobile energy conversion/storage systems applications. The electric utility industry, sponsored by U.S. Department of Energy, Electric Power Research Institute, and other funding agencies, are engaged in the development and field test demonstration of a 4.8 MW fuel cell power generating unit in downtown New York City. Successful field tests can lead to commercialization of the first generation 26 MW unit. The first generation phosphoric acid fuel cell will use liquid fuels such as naphtha and produce hydrogen via external reformers. It is estimated that in excess of 300,000 gallons of naphtha will be used for a week's operation of a 26 MW unit. Concern has been expressed in some quarters regarding the undesirable logistics associated with the delivery and storage of those quantities of flammable fuels. Hydrogen, made available from central coal gasification or electrochemical process centers by pipeline, would eliminate the fuel delivery and storage problems while overriding fuel allocation constraints. It is conceivable that the capital costs for new dedicated pipeline construction would be eliminated by injection of the hydrogen into existing transmission –distribution networks. This would not only alleviate logistics problems but would offer the option of substituting a hydrogen separation subsystem for the cost, land use and energy intensive external reformers now utilized for fuel cell power generation units. Implementation of the hydrogen injection concept will require: (1) legal and institutional barriers/incentives analyses and resolutions; (2) legal and institutional considerations pertinent to revision of fuels allocations criteria; and, (3) verification and demonstration of the compatibility of hydrogen –natural gas blends with the existing transmission–distribution network.

Industrial–Residential–Commercial–Total Energy Systems. Total energy/district heat systems utilizing fuel cells are potentially attractive options that can take full advantage of cost–design–performance tradeoffs permitted by fuel cell systems. The fuel cell's unique load–following characteristics at constant efficiency allows matching to thermal and electric loads at overall efficiencies (electric and thermal) that can approach 100%. Further, the designer can trade off electrochemical efficiency through use of cheaper catalysts and electrode structures that will not only reduce cost but will also provide more available heat for space conditioning. These systems are more likely to be implemented if regulated natural gas supplies were to be supplemented by hydrogen and hydrogen could be separated from the natural gas and used for special applications.

Automotive–Mobile Applications. Hydrogen storage and delivery systems coupled to fuel cells have the potential to provide a propulsion alternative to batteries. This approach

need not compromise commonly accepted driving practice associated with range, acceleration, and rapid refueling of vehicles.
Fuel-cell-propelled electric vehicles could provide a threefold
improvement over internal combustion engine efficiency on a
normal driving cycle and equivalent to proposed battery systems. The development of lightweight hydrogen storage systems
further enhances this application potential by permitting
increase in range and acceleration capability. The availability of hydrogen in an urban center would permit the more rapid
entry of pollution-free hydrogen-fueled vehicles into the energy infrastructure.

Economics. The ability of the fuel cell to impact the
cost of substation electricity production, total energy systems
or transportation applications is in direct proportion to fuel
utilization efficiency improvement over the competing alternatives, assuming similar capital costs. For example, a fuel--
cell-driven vehicle could be up to three times more efficient
than conventional internal combustion systems and fuel costs
per mile would be reduced proportionately. Also, a dispersed
electric power generation option available to a utility consistent with peaking demands is the gas turbine. Gas turbine
capital costs are in the range of $150-$175/kW with an efficiency on the order of 25-30%. State-of-the-art fuel cells
offer an overall efficiency approaching 40%. Figure 1 provides
cost projections for fuel cell systems indicating that large
markets will result in capital investment as low as $385/kW.
It is apparent that at the fuel cell capital costs projected
and equivalent fuel prices, substantial return on investment
can be realized.   The fuel cell offers the promise of at-
taining high-efficiency energy conversion with the added societal benefits associated with pollution-free performance. Hydrogen can be uniquely exploited by fuel cell technology such
that further development and commercialization of these conversion devices will be a central element in future hydrogen markets.

International Cooperation on Hydrogen Programs

Present efforts on hydrogen programs in some western European countries are greater than in the United States. These
programs encompass work on hydrogen production by electrochemical and thermochemical means and use in transportation vehicles, as well as supporting work on pipeline transmission and
distribution. Projects are generally a mix of long term and
near term technologies for applications responsive to the needs
of countries which have few or no indigenous energy resources,
and are largely dependent on imported energy. The interest of
European countries and Japan in alternative fuels which can be
derived from abundant hydroelectric resources, are well aware

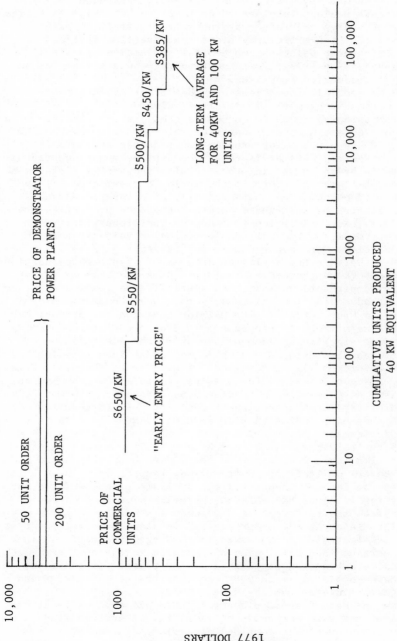

Figure 1. Power plant price schedule

that hydrogen production from such resources is an attractive
option. In Germany, work is in progress on advanced, high-
temperature nuclear reactors which have the potential to be the
source of energy to produce hydrogen by electrolytic means
coupled with high-efficiency electric conversion devices, or
with direct heat using thermochemical production processes.
The Japanese see a logical connection between solar energy and
hydrogen production processes.

Some special accomplishments that other countries have
made in the hydrogen field are cited below.

The European Economic Community (EEC), which consists of a
group of western European countries which contributes the R&D
budget and supports major programs on hydrogen at the Joint
Research Centre (JRC) at Ispra. In addition, major contractual
efforts in hydrogen are in progress with industrial firms. The
JRC has the largest effort in the world on thermochemical hydrogen production. They have recently demonstrated a bench-
scale, electrochemical-thermochemical hydrogen production process, i.e., Mark 13, which is based on the electrolysis of hydrogen bromide and thermal decomposition of sulfuric acid in a
closed cycle to produce hydrogen and oxygen.

West Germany has a similar interest in hydrogen production
and does work on theromochemical processes that can be driven
by high temperature nuclear heat sources. The leading manufac-
turer of advanced electrolyzers is the Lurgi Company, which is
located in West Germany. The Daimler-Benz Company of West Germany has major work in progress on hydrogen-fueled vehicles.
Their most significant accomplishment is the design and operation of vehicles using a dual-hydride bed of iron titanium and
magnesium nickel alloys developed originally in the U.S. Presently, discussions regarding possible cooperative programs in
this area have been begun between U.S. DOE and the German Federal Ministry for Research and Technology. In these discussions, they have indicated that a 30 man-year effort is in progress in Germany.

Both Italy and Switzerland have work in progress on advanced alkaline electrolyzers. The DeNora Corporation and
Brown Boveri of Italy and Switzerland, respectively, have built
some of the largest industrial electrolysis plants and are each
supporting in-house R&D efforts, as well as working on contracts from their respective governments.

The French have a major program in place on development of
advanced electrolytic processes for hydrogen production. They
see important industrial uses of hydrogen in the near term and
eventually a hydrogen pipeline grid in France. In addition,
they have completed a study of the feasibility of underground
storage of hydrogen, which is favorable.

In the Netherlands, pioneering work was done on metal hydrides of the rare earths, such as $LaNi_5$. A systems study was
done which showed that combined electric and hydrogen

systems may be optimal in the Netherlands. Although there are
no intensive R&D programs on hydrogen in the Netherlands, they
participate in the International Energy Agency (IEA) effort in
electrolytic hydrogen production.

In Japan, the "Sunshine Project" is a broad-base hydrogen
research program. Emphasis is on solar-assisted hydrogen production. They have done pioneering work on direct photoelec-
trolytic production from sunlight with semiconductors which has
stimulated work in the U.S.

Advanced electrolyzer development is under way in Canada
in a joint effort between Electrolyzer Corporation of America,
Noranda Corporation and the Canadian Government. Large unde-
veloped hydroelectric resources (approx. 65,000 to 100,000 megawatts) exist in Canada which could be developed quickly in the
medium term. There is good reason to believe that hydrogen
produced by a portion of the energy from this remote renewable
resource can be transported and sold to the U.S. more effectively than can extra high voltage (EHV) electric power. Evidence for hydrogen markets in the U.S. could encourage development of these hydroelectric resources in Canada.

In Brazil, with its vast remote hydroelectric resources in
the Amazon, programs are in progress to develop electrolyzer
technology and ultimately an electrolytic equipment industry.
The Gas Company of Rio, which today delivers a 50-50 vol. %
mixture of hydrogen and carbon monoxide (H_2-CO), has interest
in adding electrolytic hydrogen to its city-gas distribution
system. They presently produce the H_2-CO mixture by reforming
naphtha imported from the Middle East. Discussion between U.S.
DOE and the Brazilian Government on possible cooperative programs was initiated, but further discussions have been delayed
pending resolution of various political issues.

R&D Incentives and Requirements

In summary, one may equate interest and commitment pertinent to hydrogen technology development to the availability (or
lack) of special indigenous energy resources in a nation. The
major thrusts in hydrogen-related R&D is most evident in the
resource-starved areas of the world, such as western Europe and
Japan. Worldwide interest is brought about by recognition of
hydrogen as a future "insurance policy" and a present opportunity to exploit hydrogen's flexibility as both a chemical commodity and nonpolluting fuel supplement as well as an efficient
energy-carrying alternative to conventional·and energy wasteful
transport systems.

Major inroads to the U.S. and world energy economy will be
made by continuing to upgrade the state-of-the-art of key technologies, such as:

1. Continued development of low cost, high efficiency hydrogen production systems in areas of water electrolysis, coal

conversion, and assessing the feasibility of thermochemical decomposition processes.

2. Accelerated investigations of renewable resource conversion to hydrogen as a chemical commodity in the near term and a fuel in the longer term.

3. Development and pilot-scale testing of safe, low cost, high-capacity, hydrogen storage subsystems tied to resource conversion and resource recover options.

4. Development, test and demonstration of hydrogen end uses such as traction applications, dispersed power generation and total energy systems—market projections and development.

5. Development and/or verification of the compatibility of hydrogen as a substitute for conventional fuels within the present energy infrastructure.

Acknowledgment

This work was performed under the auspices of the U.S. Department of Energy.

RECEIVED July 12, 1979.

INDUSTRIAL TECHNOLOGY AND ECONOMICS: PRESENT AND FUTURE

Hydrogen for Ammonia Production and the Economics of Alternate Feedstocks

T. A. CZUPPON and L. J. BUIVIDAS

Pullman Kellogg, 3 Greenway Plaza East, Houston, TX 77046

One of the most important uses of hydrogen today is for the fixation of atmospheric nitrogen into the form of ammonia. While the synthesis of ammonia is principally the same in all industrial processes, the characteristics of the feedstock influence the type of process used for hydrogen generation. Steam reforming of natural gas is currently the most economic process for hydrogen manufacturing. However, the rising cost and diminishing supplies of natural gas have directed more and more attention toward the use of alternate feedstocks in recent years.

This paper analyzes the sources of hydrogen for ammonia production, presents the feed and fuel requirements of the natural gas steam reforming process, estimates the relative economics of alternate feedstocks and briefly discusses the outlook for the ammonia industry.

Introduction

One of the most important uses of hydrogen today is for the fixation of atmospheric nitrogen into the form of ammonia. Ammonia then carries, either directly or in combination with other compounds, such as urea, ammonium nitrate and various NPK materials, the supplemental nitrogen nutrient to the soil for plant growth.

Ammonia is manufactured in two basic steps, as shown in Figure 1.

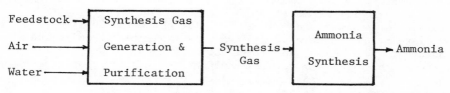

Figure 1. *Two basic steps to ammonia*

0-8412-0522-1/80/47-116-047$05.00
© 1980 American Chemical Society

First, a hydrocarbon or a carboneous feedstock, water and air are converted to synthesis gas consisting of hydrogen and nitrogen in a 3 to 1 volumetric ratio. The second step in the process is ammonia synthesis according to the following well-known reaction:

$$3H_2 + N_2 \rightleftharpoons 2NH_3$$

The catalytic conversion of hydrogen and nitrogen to ammonia is basic to all processes, while the synthesis gas generation step (hydrogen generation) varies considerably depending on the type of raw material used.

Raw Materials And Processes For Hydrogen Generation

There are basically three processes in usage today for the production of hydrogen or ammonia synthesis gas: Steam Reforming for the conversion of light hydrocarbons from natural gas to straight run naphthas; Partial Oxidation for heavy hydrocarbons and coal; and Electrolysis of Water.

Table 1 presents the main reactions and the heat requirements for the conversion of major feedstocks to hydrogen. As can be seen, in all cases, water plays a significant role as a source of hydrogen. In fact, the amount of hydrogen obtained from water increases as the carbon-to-hydrogen ratio of the feed increases; and, all of the hydrogen is obtained from water in the electrolysis process. The energy required per unit of hydrogen produced also increases in going from natural gas to water electrolysis, which tends to indicate that the preference for producing hydrogen should be directed toward low molecular weight hydrocarbons. Not only that energy requirements go up as the carbon-to-hydrogen ratio increases, but also the difficulty and the cost of processing.

Figure 2 illustrates the increased complexity in processing raw materials by the three basic synthesis gas generation techniques. While the investment is slightly lower for the ammonia synthesis step in the cases of partial oxidation of resids, coal and water electrolysis because of the high purity synthesis gas, the synthesis gas generation portion of the plant becomes more complex. An air separation plant is required for partial oxidation processes, normally the gasification step is multi-train, the equipment for carbon monoxide shift and CO_2 removal are larger, and additional equipment is required for pollution abatement. While the capital investments for water electrolysis is comparable to steam reforming plants at ammonia capacities about 300 STPD, the investment for the electrolysis route goes up dramatically for larger capacities.[1] This is because the economies of scale are only limited to the ammonia synthesis loop, while the electrolysis section investment increases almost linearly. For water electrolysis the required nitrogen can

TABLE 1.

MAJOR FEEDSTOCKS CONVERSION TO H_2

FEEDSTOCK	PROCESS	REACTION	HEAT OF REACTION AT 77°F		% H_2 FROM WATER
			BTU/ MOL FEED	BTU/ MOL H_2	
Methane (Natural Gas)	Steam Reforming	$CH_4(g) + 2H_2O(l) \rightarrow CO_2(g) + 4H_2(g)$	188,760	27,190	50
Butane (LPG)	Steam Reforming	$C_4H_{10}(g) + 8H_2O(l) \rightarrow 4CO_2(g) + 13H_2(g)$	369,920	28,455	61.5
Heptane (Naphtha)	Steam Reforming	$C_7H_{16}(l) + 14H_2O(l) \rightarrow 7CO_2(g) + 22H_2(g)$	632,630	28,756	63.6
Coal	Coal Gasification	$C_{(s)} + 2H_2O(l) \rightarrow CO_2 + 2H_2$	76,600	38,300	100
H_2O	Electrolysis	$H_2O(l) \rightarrow \frac{1}{2}O_2(g) + H_2(g)$	122,890	122,890	100

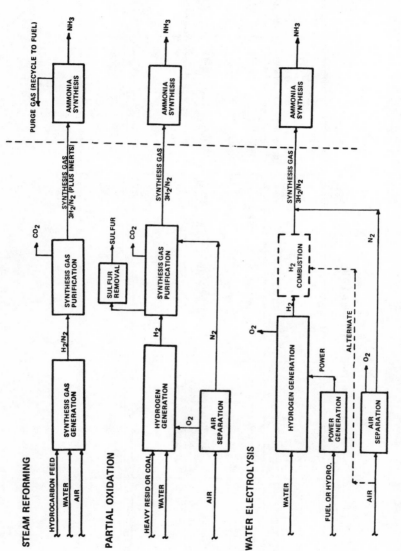

Figure 2.   Simplified ammonia-processing routes

either be obtained through air separation or air can be used
directly through a combustion step to eliminate the oxygen.
However, the disadvantage of the latter scheme is that a portion
of the hydrogen generated through electrolysis is lost in the
combustion step.

Therefore, it is not surprising that currently about 70% of
the world ammonia capacity is based on obtaining hydrogen from
natural gas reforming. Of course, the economic analysis of
various raw materials used for ammonia production is not complete
without considering their relative price and availability which
could ultimately dictate the choice for a particular feed.
Also at times, political factors enter into the choosing of a
particular feedstock for ammonia production. Table 2 shows the
approximate breakdown of the current world ammonia capacity
according to the feedstocks used. The data were derived from
SRI's "World Nitrogen" report.[2]

Table 2.
World Ammonia Capacity According To Feedstocks

Feedstock	% Of NH_3 Capacity-1978
Natural Gas	69
Naphtha	13
Coke Oven Gas	10
Fuel Oil	4
Coal	2
Refinery Gas, etc.	1.5
Electrolysis (direct or by-product H_2)	0.5
TOTAL	100

About one-half of the world's ammonia capacity based on
coke-oven gas is located in the Peoples Republic of China. In
most cases, the plants using coal, off-gases and water electrolysis are at locations where special circumstances dictate
the economic or political viability of the project. They also
tend to serve local rather than international markets.

Since natural gas plays such an important role in the
current world ammonia production, let us review the steam
reforming of natural gas process in more detail.

Ammonia By Steam Reforming Of Natural Gas

Ammonia manufacture from natural gas by the steam reforming
process is well documented. Briefly, raw synthesis gas consisting mainly of H_2, N_2, and CO_2 is produced by primary and
secondary steam reforming and CO shift conversion; this is
followed by bulk CO_2 removal, elimination of residual CO and CO_2
through methanation, and ammonia synthesis. These basic steps

in the process are illustrated in the process block flow diagram,
Figure 3 and in more detail in Figure 4.

Process Feed Requirement. The theoratical hydrogen required
for ammonia production can be calculated from the synthesis
reaction:

$$3H_2 + N_2 \rightarrow 2NH_3$$

Theoretical H_2 = 66,850 SCF/ST NH_3

The hydrogen produced in the steam reforming process is derived
from the feedstock as well as from steam. This is shown by the
following overall reaction for paraffinic hydrocarbons:

$$C_nH_{(2n+2)} + 2nH_2O \rightarrow nCO_2 + (3n + 1) H_2$$

Assuming the feedstock is methane, which is the major component of
natural gas, the theoretical feed requirement would be equivalent
to one-fourth of the potential hydrogen production or 16,713 SCF
CH_4/ST NH_3(15.2 MM BTU/ST). However, the actual process consumes
on the order of 22,420 SCF CH_4/ST NH_3 or about 20.4 MM BTU/ST NH_3
(LHV). The required quantity of feed depends on the process
design criteria chosen for the methane conversion in the
reforming section, the efficiency of CO conversion, degree of CO_2
removal and the inerts (CH_4 + Ar) level maintained in the ammonia
synthesis loop. Thus, the potential hydrogen conversion efficiency of the feedstock in the steam reforming process is about
75%. Table 3 shows where the balance of the feed is consumed or
lost from the process.

Table 3.
Feed Loss In Steam Reforming Process

Item	Type of Loss	% Of Potential H_2
Secondary Reforming	Combustion of H_2,CH_4 with O_2	15
Purge from Syn-Loop & Other Losses	Loss of H_2,N_2 and NH_3	10.5
		25.5

Actually, in the steam reforming of methane a more accurate
representation of the overall reaction taking place would be as
follows:

$$CH_4 + 1.3912 H_2O + 0.3044 O_2 + 1.1304 N_2 \rightarrow CO_2 + 2.2608 NH_3$$
$$\Delta H° \ 100°F \ = \ 36,190 \ BTU/Mol \ CH_4$$

Theoretical Feed Consumption = 17.9 MM BTU/ST

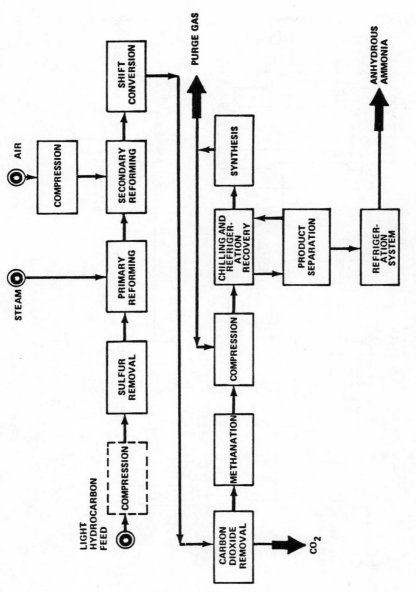

Figure 3. Ammonia by steam reforming

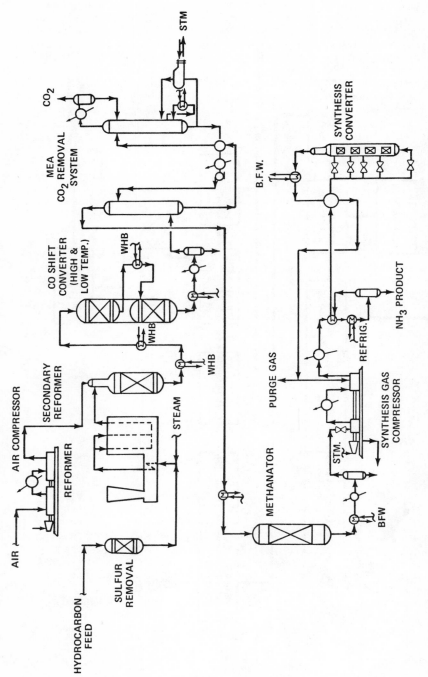

Figure 4. Flow sheet for high-capacity, single-train ammonia plant

The secondary reforming step consumes potential hydrogen by combustion with oxygen introduced in the process air. The heat of reaction, however, is used to elevate the temperature of the primary reformer effluent high enough to result in a very low residual methane in the secondary reformer effluent. Although the secondary reforming step consumes a portion of the feed through combustion, the overall effect of using the primary-secondary combination is reduced capital and operating costs.

The requirement for purging synthesis gas from the ammonia synthesis loop is due to incomplete conversion of the feedstock to H_2 and the production of CH_4 in the methanator (residual methane), and to the introduction of inert argon with the process air. Both residual methane and the argon act as inerts in the ammonia synthesis and their concentration would build up in the loop if not purged. Although the purge gas is a process material loss and proportionally increases the feed requirement, it is recovered for its fuel value in the primary reformer. Several processes are commercially available to recover principally hydrogen from the ammonia plant purge gas through adsorption or cryogenic techniques. The hydrogen is then recycled back into the process for either reducing the feed and fuel requirement or produce additional ammonia. Pullman Kellogg recently developed a system that converts the purge gas into additional ammonia which can be applied either for incremental ammonia production or to reduce the plant's energy requirement without increasing overall production rate.

Fuel Requirement. Most of the fuel used in a steam reforming plant is required to sustain the following highly endothermic reaction in the primary reforming furnace:

$$CH_4 + H_2O \rightleftharpoons CO + 3H_2 \qquad \Delta H^\circ = 88,660 \text{ BTU/Mol } CH_4$$

This is the methane-steam reaction and it occurs at a high pressure (450-500 psig) and a temperature of about 1500°F. Additional energy is required to run the process air, synthesis gas and refrigeration compressors; however, most of this is obtained from waste heat contained in the furnace flue and process gases. There is a very high degree of heat recovery in the process. In fact, 75% or more of the total plant energy (excluding feed) requirement is obtained from waste heat. A large part of the process waste heat is used to generate high pressure steam, which provides the motive power for the compressor turbines, and the reforming steam. Normally, an integrated auxiliary boiler is installed to provide an energy-balanced plant. Table 4 summarizes the feed and fuel requirements in current ammonia plants based on natural gas feedstock.

Table 4.
Current Ammonia Plant Feed And Fuel Requirements

	MM BTU/ST (LHV)	%
Natural Gas Feed	20.4	65.8
Natural Gas Fuel (Net)	10.6	34.2
TOTAL	31.0	100 %

Economics of Alternate Feedstocks

In evaluating the present and future technology for the
production of ammonia, capital and energy costs are keys to
making a process selection. Since the emergence of the energy
crises, numerous publications have studied the economics of
alternate feedstocks for ammonia.[1,3,4,5,6,7] While the absolute
investment figures for the alternate feedstocks vary significantly in the publications, there appears to be a general agreement on their relative values. The relative investments and
energy requirements shown in Figures 5 and 6 respectively, tend
to agree with most of the recently published figures. Thus,
the capital cost of a fuel oil-based ammonia plant appears to be
about 1.5 times and for a coal-based plant about 2.0 times that
of an ammonia plant using natural gas feedstock. Besides the
difficulty in arriving at accurate investment figures, there are
a number of other factors which play important roles in deter-
mining which feedstock to use. Among these are the availability
of feed and fuel, price structures, plant size and location,
transportation cost, contractual consideration, cost of capital
and international market situation. Thus, one can only speculate
as to how a few of these factors will affect the choice of feedstock which will briefly be discussed with emphasis on the unit
price of feed.

Table 5 presents the estimated ammonia production costs
using current feedstock prices. Also, Figure 7 was prepared to
show the effect of feed and fuel price on ammonia manufacturing
costs. The investment figures for the natural gas and naphtha
feedstocks are representative of plants instantaneously erected
on the U.S. Gulf Coast in early 1979 and include the minimum off-
site facilities, spare parts and product storage. Also, utilities are assumed to be available at the plant gate. Investment
required for the development of infrastructure, feedstock
delivery and product distribution, interest during construction
and forward escalation are excluded. The investments shown for
the partial oxidation of fuel oil and coal gasification were
based on the above mentioned factors relative to the natural
gas-based plant. The production costs of ammonia developed in
Table 5 are detailed enough to enable the reader to insert other
price structures which are suitable to his own situation.

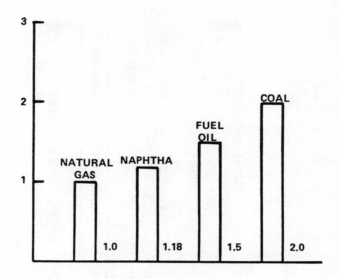

Figure 5.   Relative NH₃ plant investment vs. feedstock (1150 STPD)

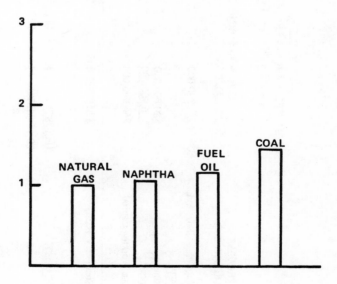

Figure 6.   Relative NH₃ plant energy use vs. feedstock (1150 STPD)

TABLE 5.
ESTIMATED AMMONIA PRODUCTION COSTS
1150 STPD CAPACITY

PROCESS FEEDSTOCK	STEAM REFORMING		PARTIAL OXIDATION (TEXACO)	
	NATURAL GAS	NAPHTHA	HEAVY FUEL OIL	COAL (K-T)
Annual Production – short tons	380,000	380,000	380,000	380,000
INVESTMENT				
Fixed Investment	$60,000,000	$71,000,000	$90,000,000	$120,000,000
Working Capital[1]	6,900,000	10,300,000	9,900,000	8,500,000
Total Fixed & Working Capital	$66,900,000	$81,300,000	$99,900,000	$128,500,000
Feedstock Cost (LHV Basis)[2]	$2.50/MMBTU	$3.50/MMBTU	$2.50/MMBTU	$1.00/MMBTU
OPERATING COST	Units/ST $/ST	Units/ST $/ST	Units/ST $/ST	Units/ST $/ST
Feedstock	20.4 MMBTU 51.00	20.4 MMBTU 71.40	26.9 MMBTU 67.25	33.1 MMBTU 33.1

	Plant 1		Plant 2		Plant 3		Plant 4	
Utilities								
Fuel (same as feed)	10.6 MMBTU	26.50	12.1 MMBTU	42.35	7.6 MMBTU	19.00	11.9 MMBTU	11.9
Power @ $0.03/kwh	15.5 kwh	0.47	21.6 kwh	0.65	–	–	–	–
Water @ $0.05/M Gal.	2400 Gal.	0.12	2500 Gal.	0.13	2700 Gal.	0.14	3835 Gal.	0.19
Catalyst & Chemicals	–	1.20	–	1.30	–	0.68	–	0.68
Total Feed & Utilities		79.29		115.83		87.07		45.87
Labor @ $18,000/manyear	30 men total	1.42	30 men total	1.42	37 men total	1.75	75 men total	3.55
Labor & Plant Overhead	100% of Labor	1.42	100% of Labor	1.42	100% of Labor	1.75	100% of Labor	3.55
Interest on Working Capital	10%/yr	1.82	10%/yr	2.71	10%/yr	2.61	10%/yr	2.24
Indirect Charges[3] @ 18% of Fixed Capital/Year		28.42		33.63		42.63		56.84
Total Production Cost		112.37		154.01		135.81		112.05
Pretax Return @ 20% Fixed Capital/Year		31.58		37.37		47.37		63.16
Cost of NH$_3$ F.O.B. Plant		143.95		191.38		183.18		175.21

NOTES:
(1) Working capital = 1 month feedstock supply @ cost, except gaseous feed & 2 months ammonia storage @ $100/ST.
(2) LHV naphtha = 18,900 BTU/Lb, Fuel oil = 18,380 BTU/Lb, Coal = 10,780 BTU/Lb.
(3) Indirect Charges: 10% Depreciation, 2% Insurance, 4% Maintenance (Labor & Material), 2% General Administrative.

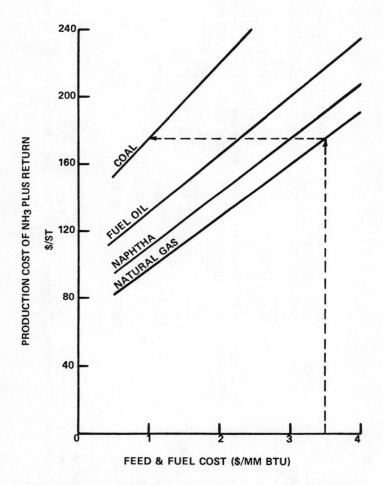

Figure 7. Estimated ammonia production cost vs. feed and fuel cost

Natural Gas Feed. A high efficiency operation was assumed including use of a 1500 Lb./Sq. inch gauge steam system. In addition, this design includes many process features for increasing thermal efficiency including combustion air preheat for the reformer, Union Carbide's UCAR Amine Guard for the CO_2 removal system, a hydraulic turbine for power recovery in the CO_2 removal system, additional shift catalyst to reduce residual CO and synthesis loop purge, an additional stage of refrigeration in the loop to reduce compressor horsepower, and efficient steam turbines coupled with much higher steam superheat temperatures.

Naphtha Feed. A naphtha-based plant presently costs about 18% more than a natural gas-based ammonia plant[3].Furthermore, naphtha feeds require more utility in terms of CO_2 removal, reforming steam, and feed pretreatment. Also, the unit cost of feed is usually much greater than natural gas on a calorific basis. Depending on price structure, the operating cost can vary over a wide range. If lower grade fuels can be used for firing the reformer, reductions in operating cost can be realized bearing in mind that maintenance increases with use of lower grade fuels. For example, the potential cost saving with the use of a lower grade fuel at $2.50/MM BTU (LHV) would be $12.10/ST.

Heavy Fuel Oil. A partial oxidation unit based on heavy fuel oil will cost significantly more than a natural gas-based plant because it requires an air separation plant, additional desulfurization, carbon removal and recycle, and other facilities. Nevertheless, if the cost of fuel oil is sufficiently low, one could justify a design selection based on partial oxidation. In view of the uncertainty of equipment cost information, it is difficult to pinpoint investment differences between a steam reformer plant and a partial oxidation unit. Estimates are that the investment difference varies over a wide range. Some recent reports reveal that the difference may be 50% and possibly higher.[5,6] Direct operating data for the Texaco partial oxidation process were taken from Child and Marion[8] paper. The plant, equipped with all steam drives, incorporates 1200 psig gas generation, shift conversion by sulfur-active CoMo catalyst and acid-gas removal by the Rectisol process.

Coal Feed. As in the case of steam reforming versus partial oxidation, a plant based on use of coal as feedstock is expected to cost more than one based on partial oxidation because of additional equipment requirements, particularly with regard to coal handling operations, gasification and raw gas treatment. Despite these cost differences, coal based plants can be justified if the cost of coal is sufficiently low, that is, relative to light or heavy hydrocarbons. The approximate investment for a coal-based plant appears to be about 2.0 times that for a natural gas plant.[5,6,7] The investment is about 33% more for a fuel oil

plant. Direct operating data for the ammonia from coal process
were taken from Staege[9] paper for the Koppers-Totzek process.

Comparing partial oxidation of fuel oil with coal gasification differs somewhat from comparisons with steam reforming.
Both fuel oil and coal processing involve the partial oxidation
techniques and similar downstream gas treating schemes. Furthermore, both involve use of an air separation plant and sulfur
recovery facilities. Differential investment for the two
designs is not as great as that reported for the reforming vs.
coal gasification comparison, and thus a more careful economic
appraisal would be required which would have to include such
factors as fuel oil availability over a lengthy period, labor
costs covering coal mining operations, price stability, transportation costs for both product and feed, and plant location.

5. Coal based operations can be justified in
    locations where the unit cost of coal is low,
    where gas is not available, and where the
    alternative is expensive imported oil.
    However, the use of such feed will be
    governed to a great degree by whether an
    attractive price fifferential between solid
    and hydrocarbon feedstocks can be maintained
    on a long term basis. The high costs of
    mining and transportation charges will, of
    course, reduce any potential economic
    advantage inherent in use of solid feeds.

Outlook For The Ammonia Industry

A critical factor in the manufacture of ammonia today is
its energy consumption. The rising energy costs in the U.S.
and elsewhere have directed more and more attention to improving
the plant's energy requirement. Programs for energy reduction
have been or are being pursued in several areas. Opportunities
for energy savings in existing ammonia plants have been explored
and some already implemented. A series of ammonia plant designs
have been developed with progressively lower energy consumptions
than those of plants now in operation. In many places, alternative, less expensive, feedstocks are being seriously considered.
Processes associated with Chemo-nuclear energy, sea water
temperature gradients, and even solar energy are mentioned as
possibilities in the very distant future.

Let us briefly review the developments of Kellogg's natural
gas-steam reforming technology with respect to energy consumption.
As shown in Figure 8, the total energy (natural gas plus electric
power) consumption has been reduced from about 48.5 MM BTU/ST
in the early 1950's to about 33 MM BTU/ST in 1965, representing
a reduction of 36%. Since 1965, some further reduction of energy
has taken place by the use of such efficiency improving facilities
as combustion air preheat for the primary reformer furnace, purge

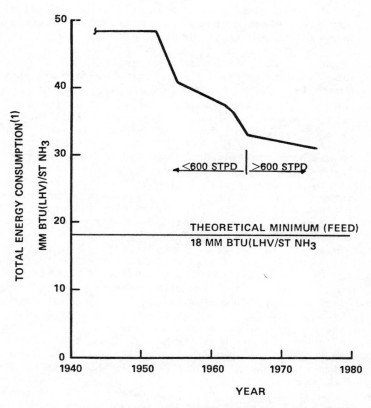

(1) TOTAL ENERGY = FEED + FUEL + ELECTRIC POWER;
     ELECTRIC POWER = 10,500 BTU/KWH

*Figure 8.   Energy consumption of Kellogg ammonia plant's steam reforming of
natural gas*

gas recovery, Union Carbide's UCAR   Amine Guard for the CO_2 removal system, increased number of refrigeration stages and low pressure drop equipment (e.g. horizontal ammonia converter), which will collectively result in an energy consumption of about 30 to 31 MM BTU/ST.

The diminishing supply of inexpensive energy has again brought about an acceleration of R & D efforts aimed at further reducing the energy requirements. One needs to bear in mind, however, that the minimum energy requirement of an ammonia plant based on natural gas feed is about 18 MM BTU (LHV)/ST and this is equivalent to the theoretical quantity of feedstock (CH_4)) that cannot be lowered. While it becomes increasingly difficult to design plants approaching this theoretical minimum energy consumption, there is room for improvement to the currently operating units. Many of the energy improving techniques considered today are merely extensions of established technology and have been known, or even practiced in part, for some time in the past.

As ammonia plant designs are developed with progressively lower energy consumption, the material and energy balances become more closely integrated and the interaction of the process and utility systems become more complex. The type of plant design will be dictated much more than in the past by the project particulars, especially by the feed and utility price structure. Due to the wide variations of feedstock prices at various locations in the world, it is no longer feasible to apply the so-called "standard" plant design approach for all situations. Thus, plants will be built based on proven, well established technologies in areas where feedstock prices are not of a major concern. Improved technologies with respect to energy consumption will be of prime importance in areas where energy costs are high. As the energy consumption of the ammonia plant is improved, the capital investment tends to increase. Thus, each improvement to the current plant design must meet an economic payout period for an optimum plant design, which is highly dependent on the price of feed and fuel.

In the short range, we can expect increased use of the already established energy saving devices mentioned before as well as additional ones. These are summarized below:

1.   Improved ammonia synthesis, e.g. horizontal ammonia converter.

2.   Purge gas recovery.

3.   Combustion air preheat.

4.   More heat recovery.

5. High efficiency steam turbines and compressors.

6. Improved reformer tube metallurgy to allow more optimum reforming conditions.

7. Reduced pressure drops.

8. More efficient CO_2 removal systems, e.g. physical solvents.

9. Improved energy cycle, e.g. use of gas turbines.

10. Higher activity catalysts.

Incorporating some or all of the above mentioned improvements will lead to a considerable reduction of energy consumption of an ammonia plant based on steam reforming of natural gas or naphtha. The actual plant design will depend on the project particulars and feed and utility price structure.

At this time, it is worth mentioning that improvements in ammonia technology are not limited to steam reformer-based plants. Advancements have been made in synthesis gas generation via the non-catalytic partial oxidation processes offered by Texaco and Shell, particularly toward applying their well established technologies in fuel oil partial oxidation to coal gasification.[7,10,11] Coal gasification technologies are also undergoing very intensive development programs.[11,12] Interest is again directed toward the classic water electrolysis route for generating hydrogen.[1,13]

Thus, as technologies develop in all these areas of ammonia synthesis gas generation, there will be a new set of guidelines for analyzing the relative economics of alternate feedstocks.

APPENDIX

Abbreviations and Definitions

```
kwh = 10,000 Btu
lb. = pound
LHV = lower or net heating value
          CH₄ = 909.9 Btu/scf
          H₂  = 274.5 Btu/scf
 MM = million
scf = standard cubic feet @ 60°F 14.7 psia
 ST = short ton
STPD = short ton per day
```

REFERENCES

1. "Electrolytic Hydrogen for Ammonia Synthesis", The British Sulphur Corporation, Ltd., *Nitrogen*, No. 97, September/October 1975, P. 39.

2. "World Nitrogen - 1976 Edition", Volume II: Plant Listing, Stanford Research Institute, Menlo Park, California.

3. Buividas, L. J.; Finneran, J. A.; Quartulli, O. J., "Alternate Ammonia Feedstock", Chemical Engineering Progress, October 1974.

4. "Outlook for Ammonia Industry", Process Economic Reviews, Report No. PEP '75-2-2, Stanford Research Institute, Menlo Park, California, December 1975.

5. Appl, Dr. M., "A Brief History of Ammonia Production From The Early Days To The Present", Nitrogen, 100, 47-58 (1976).

6. "Production Economics For Hydrogen, Ammonia And Methanol During The 1980-2000 Period, Exxon Research and Engineering Company, Government Research Contract No. 368150-S April 1977, Appendix A by Chem Systems Inc.", Cost of Production Estimates For Hydrogen, Ammonia And Methanol, October 21, 1976, P. 127-166.

7. Waitzman, D. A., "A Technical And Economic Review Of Coal-Based Ammonia Production", Proceedings of The 27th Annual Meeting Fertilizer Industry Round Table, 1977, P. 33-41.

8. Child, E. T.; Marion, C. P., "Recent Developments In The Texaco Synthesis Gas GEneration Process", Presented at The Fertilizer Association Of India National Seminar, December 14-15, 1973, New Delhi.

9. Staege, Hermann, "The Production Of Intermediate And Final Products In Synthesis Gas Chemistry According To The Koppers-Totzek Process, Using Coal Gas", Chemical Age of India, Volume 29, No. 12-A, December 1978.

10. European Chemical News, April 15, 1977, P. 30.

11. "Coal Takes Aim At Synthesis Gas", Chemical Week, October 5, 1977, P. 34.

12. Schora, F. C.; Blair, W. A., "The Status Of Coal Gasification, 1977" IGT Paper presented at the Sixth Airapt International High-Pressure Conference, Boulder, Colorado July 24-25, 1977.

13. Nuttall, L. J., "Water Electrolysis Using Solid Polymer Electrolytes", General Electric Co., Presented at the Institute of Gas Technology's "Hydrogen for Energy Distribution Symposium", July 24-28, 1978.

RECEIVED July 12, 1979.

Hydrogen in Oil Refinery Operations

HAMPTON G. CORNEIL and FRED J. HEINZELMANN

Exxon Enterprises Inc., P.O. Box 192, Florham Park, NJ 07932

A study (Reference 1) completed in April, 1977, was conducted by Exxon Research and Engineering Company for the Conservation Division, Department of Energy, to predict the quantities of industrial hydrogen that will be needed in the U. S. during the 1980-2000 period for the current industrial uses and to determine the costs of producing these industrial hydrogen products by several alternative processes that are likely to be used commercially or considered for commercial use during this period. The data concerning crude runs to stills and hydrogen processing have been updated in this study to reflect current estimates of these values. Hydrogen for use in petroleum refining operations provides a major hydrogen requirement today and this will continue throughout the 1980-2000 period. This paper discusses the uses of hydrogen in petroleum refining, the quantities of hydrogen that will be required during the 1980-2000 period, and the economics of producing this hydrogen.

As is explained in Reference 1, investment and operating cost data for various hydrogen manufacturing processes were provided for this study by Chem Systems, Inc., New York, New York. These economics data were developed using identical methods and assumptions, thereby permitting side-by-side comparisons of the several processes.

Conclusions

Hydrogen willfully produced for use in U. S. refineries for gas oil desulfurization and residuum hydrocracking and desulfurization probably will increase from about 1.8 billion SCF/D in 1980 to 3.9 billion SCF/D in the year 2000. These figures

0-8412-0522-1/80/47-116-067$07.00
© 1980 American Chemical Society

exclude refinery by-product hydrogen (from octane improvement reformers) which is used for desulfurizing various light distillate products.

Steam reforming using natural gas and other light hydrocarbon feed stocks will continue to be the most attractive method of manufacturing willfully-produced refinery hydrogen. Feed stocks required for reforming in U. S. refineries will increase from about 150,000 B/D crude oil equivalent in 1980 to 330,000 B/D in the year 2000.

Partial oxidation of resid feed stocks is used for producing 5% to 7% of current willfully produced hydrogen. Its use will not increase significantly because manufacturing costs are, in general, higher than with steam reforming.

Coal gasification probably will not be used to manufacture refinery hydrogen in the U. S. to any significant extent during the 1980-2000 period. Capital investments and operating costs for coal gasification in the four major refining centers are likely to be much higher for coal gasification than for steam reforming.

Hydrogen Uses in Petroleum Refining

Prior to the 1965-1970 period, most of the hydrogen used in petroleum refining was used for treating light naphthas and middle distillates to provide for desulfurization and product stability. These hydrogen-treating operations require little hydrogen, ranging from 10-20 SCF/B for the light naphthas to 100-200 SCF/B for the middle distillates. Stringent environmental restrictions requiring lower sulfur content products have accelerated the use of these hydrogen-processing operations. Hydrogen for these operations has in the past and will continue to be provided as by-product hydrogen produced in naphtha reformers operated to increase the octane number of motor gasoline fractions. Although this by-product hydrogen is typically of only 70% to 80% hydrogen content, this low-purity hydrogen is satisfactory for hydrogen-treating these light distillates.

In recent years, a more severe type of hydrogen treating has been added to refinery processing systems in which heavy distillates (gas oil) and residuum are hydrotreated to remove sulfur and to convert these heavier hydrocarbons to products of lower molecular weight. The addition of "external" hydrogen increases the H/C ratio of the products substantially above that of the feed stocks. The use of these hydrocracking and hydrodesulfurization processes has become increasingly important as crude oil supplies have become heavier (of lower H/C ratio) and of higher sulfur content and because of the economic need to convert residuum to distillate products.

The conversion of residuum to lighter products can also be conducted with thermal cracking (vis breaking) or coking, a process that removes carbon from the system, thereby increasing the H/C ratio of the coker products relative to that of the feed stock. Figure 1 is a simplified flow diagram of refinery processing. In this system, conversion of resid (650°F +) is provided by either hydrotreating or coking. Figure 2 shows the system assuming hydrotreating is used for resid conversion. The processing sequence shown in Figure 3 assumes coking is used for residuum conversion.

The choice between these two alternative residuum-processing methods depends on many economic factors, and how such a choice would be made is beyond the scope of this study. However, these two processes are compared in Reference 2. Figure 4, a summary chart from Reference 2, indicates that coking is the preferred choice if it is desired to convert 60% or more of the crude (Arabian Heavy, in this case) to prime (clean) products.

When hydrogen is used for hydrodesulfurizing and hydrocracking of gas oil (heavy distillates) and residuum, large quantities of high-purity hydrogen (95%+) are required as inputs to these plants, and the by-product hydrogen from octane improvement reformers is not adequate. Hydrogen consumption for these processes range from 300 SCF/B for light gas oil desulfurization to 3000 SCF/B for severe hydrocracking. Since this hydrogen requirement is substantial and by-product hydrogen from other operations is not adequate, these gas oil desulfurizers and resid hydrocrackers are installed with their own hydrogen-generating facilities.

Hydrogen Requirements for Petroleum Refining

The principal current U. S. requirements for industrial hydrogen are for use in petroleum refining, for the manufacture of ammonia and methanol, and for a wide variety of small uses including chemicals manufacture, metallurgy, welding, etc. It is beyond the scope of this paper to discuss the factors that will affect the future requirements for ammonia, methanol, and small-user hydrogen; however, these factors are discussed rather completely in Reference 1.

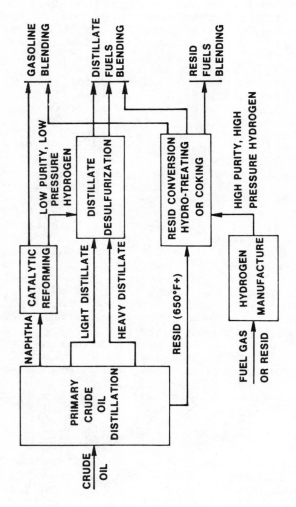

Figure 1. Simplified refinery processing

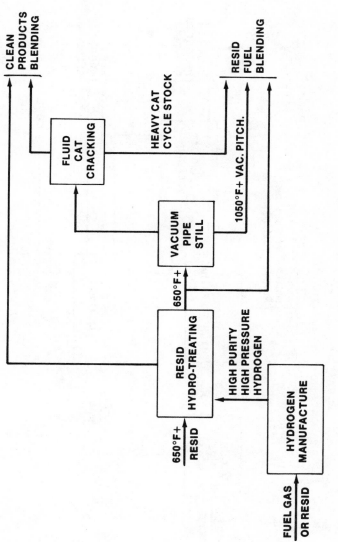

Figure 2. Resid processing by hydrotreating

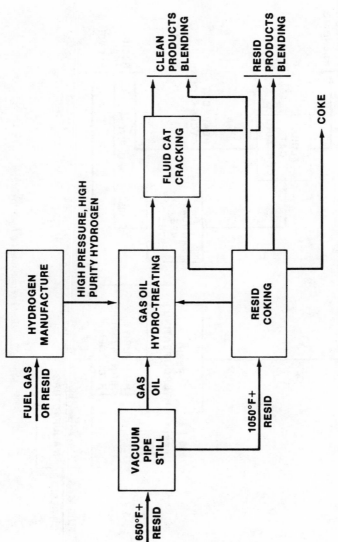

Figure 3.   Resid processing by coking

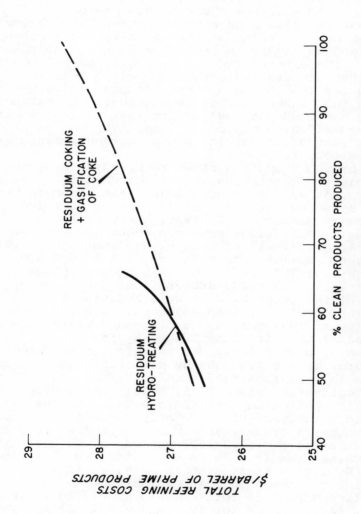

Figure 4. *Total refining costs for resid hydrotreating vs. resid coking (U.S. Gulf coast 1980—1980 dollars)*

Table 1 summarizes the future requirements for
refinery hydrogen, ammonia, methanol, and small-user
hydrogen. Refinery hydrogen requirements will be
discussed in more detail below. As has already been
noted, refinery hydrogen represents the second largest
industrial hydrogen use, second to ammonia manufacture.
The data in Table 2 show these future hydrogen
requirements expressed on a heat content basis using
the higher heating value of the refinery and small-user
hydrogen and the hydrogen contained in the ammonia and
methanol. These total requirements will be 0.78 Quad/
year in 1980 and are projected to be 1.67 Quad in the
year 2000 for the coking option. (A Quad = 1.0 x 10^{15}
Btu using the higher heat of combustion.) Using these
data, the compounded annual growth rate for these uses
during the 1980-2000 period is 3.9%. These data as
Quads per year of industrial hydrogen requirements are
shown in Figure 5.

The hydrocarbon feed stock required to produce
these projected quantities of refinery hydrogen, small-
user hydrogen, ammonia, and methanol are shown in
Table 3, assuming reforming is used to produce 100% of
each of these products. In 1975, these hydrogen
manufacturing operations consumed about 450,000 B/D
(crude oil equivalent) of natural gas and light naphtha
feed stocks. By 1980, this figure will be 632,000 B/D,
and by the year 2000, the hydrocarbon consumption is
predicted to be 1,374,000 B/D OE (coking option) or
about 6% of the total U. S. liquid hydrocarbon
requirements or 14% of the projected crude oil imports
for that year. However, as is discussed in Reference 1,
a substantial part of these future hydrogen
requirements, particularly for ammonia and methanol
manufacture, may be produced by coal gasification, thus
conserving hydrocarbon feed stocks.

The estimated future U. S. requirements of
hydrogen for petroleum refining are based on the
forecasted quantities of crude oil, natural gas liquids
(NGL), synthetic crudes, and imported crudes shown in
Table 4. During the 1980-2000 period, U. S. crude runs
will increase from 15.3 to 18.0 million B/D while the
total U. S. liquid hydrocarbon supply, including
product imports, will increase slightly from 19.7 to
21.1 million B/D. The total liquid supply corresponds
to 40 and 43 Quad/year for 1980 and 2000. By the year
2000, there will be a substantial increase in the
quantity of residuum converted due to an overall
increase in the quantities of heavy crudes in the crude
slate.

TABLE 1.

U. S. HYDROGEN REQUIREMENTS

	1975	1980	2000
Refinery Hydrogen, MSCF/CD	1220	1780	4725
Ammonia, MST/year	16.0	21.3	41.8
Methanol, MST/year	3.8	5.3	15.3
Small User Hydrogen, MSCF/CD	208	264	706

TABLE 2.

U. S. INDUSTRIAL HYDROGEN--1980-2000

Quads of H_2 Per Year

YEAR	1980	1990	2000 Coking Option	2000 Hydro-Processing Option
Refinery hydrogen	0.21	0.34	0.46	1.41
For ammonia manufactured in U. S.	0.46	0.66	0.90	0.90
For methanol " "	0.08	0.14	0.23	0.23
For all small users	0.03	0.05	0.08	0.08
Total industrial hydrogen	0.78	1.19	1.67	2.62

Note: The above data exclude new uses for industrial hydrogen that may develop during the 1980-2000 period.

TABLE 3.

PETROLEUM FEED STOCK REQUIRED ASSUMING ALL U. S.
INDUSTRIAL HYDROGEN IS PRODUCED BY STEAM REFORMING

Thousand Bbl/CD crude oil equivalent

YEAR	1980	1990	2000 Coking Option	2000 Hydro-Processing Option
For refinery hydrogen	149	242	326	1001
For ammonia	368	527	722	722
For methanol	93	166	267	267
For all small users	22	36	59	59
Total	632	971	1374	2049
Total/U. S. liquid oil, %	3.2	4.8	6.5	9.7
Total/U. S. oil imports, %	6.5	7.5	13.6	20.3

TABLE 4

U. S. LIQUID PETROLEUM SUPPLY, 1980-2000

	1975	1980	1990	2000
Total liquid supply, million B/D	16.6	19.7	20.4	21.1
Total runs to stills, " "	12.5	15.3	16.7	18.0
Total liquid supply, Quad/year	33.6	39.9	41.3	42.7

Data to show crude runs to stills, total liquid supply, and resid yield on crude are plotted in Figure 6. As was mentioned previously, the data used in this study for crude runs to stills and future hydrogen-treating conditions have been updated over those used in the 1977 study (Reference 1) to reflect the current outlook for liquid product requirements. For comparison purposes, data points shown in Figure 6 indicate the total liquid supply data used in the 1977 study for the years 1990 and 2000. The total liquid supply data used in this study are less than those used in the 1977 study by about 3.0 million B/D in 1990 and 3.5 million B/D in the year 2000.

The corresponding charge rates for the main refinery processing operations are shown in Table 5. These data show that the quantity of residuum charged to coking plus hydrotreating will increase from 1.5 million B/D in 1980 to about 3.9 million B/D in the year 2000, assuming coking is the main residuum conversion procedure (Column 3), or 6.2 million B/D if resid hydrocracking is the main resid conversion method (Column 4). The quantities of gas oil and residuum charged to hydrogen-processing operations will increase from 1.9 million B/D in 1980 to 5.9 million B/D in the year 2000, assuming coking is the main resid conversion process or 9.6 million B/D assuming hydrocracking is the main resid conversion process.

The question of whether to process the residuum by coking or hydrocracking obviously has a very substantial impact on the quantity of refinery hydrogen needed for all processing operations. The choice between these two alternative residuum processing systems depends on many factors, such as the characteristics of the crudes that must be processed, the quantity and quality of products that are needed, and many economic factors, such as crude costs, product values, etc. The data in Table 5 are based on a comparison of coking and hydrotreating using economic assumptions essentially the same as those used in Reference 2. This calculation results in the optimum processing mix for the year 2000 being that shown in the third column of Table 5 headed, Coking Option. However, as a sensitivity, the case shown in the fourth (last) column of Table 5 headed, Hydroprocessing Option was calculated to show what might occur in the year 2000 if economic factors changed to favor hydroprocessing over coking. This case would provide the maximum hydrogen requirement for the year 2000 that conceivably would occur.

TABLE 5.

PROCESSING AT U. S. REFINERIES, 1980-2000

YEAR	1980	1990	2000 Coking Option	Hydro-Processing Option
Average Annual Charge Rates, Million B/D				
Primary distillation	15.3	16.7	18.0	18.0
Catalytic cracking	5.2	5.3	5.3	5.3
Catalytic reforming	3.1	4.2	4.7	4.7
Resid coking	0.8	1.8	2.7	0.5
Gas oil desulfurization	1.2	2.7	4.7	3.9
Resid desulfurization	0.1	0.3	0.3	1.6
Resid hydrocracking	0.6	0.8	0.9	4.1
Subtotal--resid processed	1.5	2.9	3.9	6.2
Subtotal--hydrogen processed	1.9	3.8	5.9	9.6

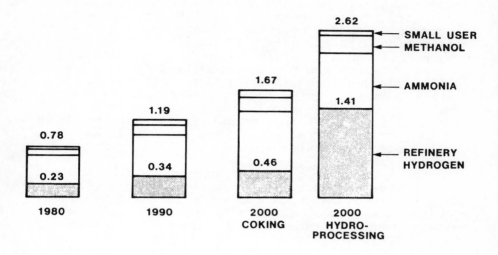

Figure 5.   *U.S. industrial hydrogen requirements, 1980–2000 (quads per year of hydrogen)*

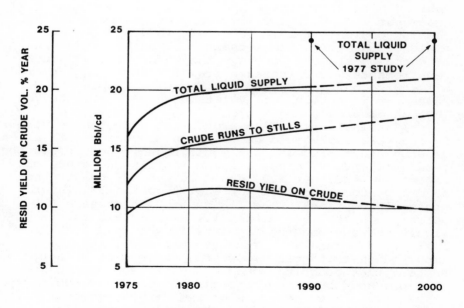

Figure 6.   *U.S. refinery crude runs, 1975–2000*

These caluclations of refinery hydrogen
requirements are based on hydrogen consumption data
summarized in Table 6 for the various processes that
consume and produce hydrogen. These hydrogen
consumption data allow for the theoretical hydrogen
consumption and 20% excess above that required to
maintain the desired hydrogen purity in the process.
If the hydrogen input to the process is produced by
steam reforming, the hydrogen not consumed by the
hydrogen-processing operation can be recovered by
adding it to the reformer feed. If the hydrogen were
provided by some alternative process, the excess from
the hydrogen-processing operation would be burned as
refinery fuel.
        The data on processing requirements from Table 5
and the hydrogen consumption data from Table 6 are the
basic variables in calculating the refinery hydrogen
requirements. These data in Table 7 indicate that
refinery hydrogen requirements will increase from 1.8
billion SCF/D in 1980 to 3.9 billion in the year 2000
(coking option). Under the conditions corresponding to
the maximum use of hydrogen processing in the year 2000
(hydroprocessing option, last column), the refinery
hydrogen requirements would be 12.0 billion SCF/D.
Table 7 also shows these hydrogen requirements as 0.21
Quad/year in 1980 and 0.46 Quad/year in the year 2000
(coking option) or 1.41 Quad/year (hydroprocessing
option). Assuming these quantities of hydrogen are
produced by steam reforming, the reformer feed stock
requirements would increase from 149 thousand B/D crude
oil equivalent in 1980 to 326 thousand B/D in the year
2000   option) or 1001 thousand B/D for the year
2000 (hydroprocessing option).
        If all the petroleum-processing hydrogen required
in the year 2000 could be supplied by coal gasification
or some other process that would require no hydrocarbon
feed stock, U. S. crude oil consumption and crude oil
imports in the year 2000 could be reduced by 326 thousand B/D (coking option) or 1001 thousand B/D (hydroprocessing option). The figure of 326 thousand B/D
corresponds to 1.5% of the 21.1 million B/D total
liquid hydrocarbon requirements in the year 2000 or
3.2% of the crude and products that are likely to be
imported in that year. However, these data do not
suggest that substitution of nonpetroleum hydrogen for
that produced by steam reforming would be economically
attractive. The economics of refinery hydrogen
production are discussed in the following sections.

TABLE 6.

PETROLEUM PROCESSING WITH HYDROGEN

PROCESS	Hydrogen Consumption (Production) SCF/Bbl	Reactor Conditions	
		°F	PSIA
Catalytic reforming	(800-1200)	500-1000	150-500
Naphtha hydrotreating	10-50	500-650	200-500
Distillate hydrotreating	100-300	550-750	250-800
Gas oil desulfurization	400-1000	650-800	1000-2000
Resid desulfurization	600-1200	650-800	1000-2000
Resid hydrocracking	1200-1600	750-800	2000-3000

TABLE 7.

HYDROGEN REQUIRED FOR U. S. PETROLEUM REFINING

YEAR	1980	1990	2000 Coking Option	2000 Hydro-Processing Option
Hydrogen required, million SCF/CD for				
Gas oil desulfurization	420	950	1650	1370
Resid desulfurization	210	390	450	2400
Resid hydrocracking	1150	1550	1800	8200
Total	1780	2890	3900	11970
Total Quad/year	0.21	0.34	0.46	1.41
Feed stock required, thousand B/D crude oil equivalent	149	242	326	1001

Capacities of plants for supplying hydrogen for use in petroleum refining are usually in the range of 50 to 100 million SCF/D. Processes now available for producing these large quantities of hydrogen are steam reforming, resid partial oxidation, and Koppers-Totzek (hereinafter called K-T) coal gasification. Several other coal gasification processes are being developed and one or more of these may prove to be attractive for use during the 1980-2000 period. These (hereinafter called new coal gasification) include a high-pressure partial oxidation process being developed by Texaco, a similar process being developed jointly by Shell and K-T, and the U-Gas and Steam-iron processes being developed by the Institute of Gas Technology. These processes are discussed in various papers, including References 1, 3, and 4.

Steam reforming, using natural gas or naphtha feed is now used to produce 90% or more of U. S. refinery hydrogen requirements. As of January, 1976, there were 32 plants in the U. S. producing hydrogen for use in petroleum refining. The capacity of these plants totaled 1445 million SCF/D. Included in this figure are two California plants employing resid partial oxidation.

Economics of Refinery Hydrogen Manufacture

Estimates have been prepared to indicate the cost of producing refinery hydrogen during the 1980-2000 period. The assumed inflation rates shown in Table 8 were used to estimate the future cost of oil, gas, and coal shown in Table 9. Inflation rates shown in Table 8 also assume construction costs will not exceed the general inflation rate. During the 1980-2000 period, the general or overall rate of inflation of all commodities including coal, electricity, and construction costs is assumed to average 5% per year.

Prices of natural gas and petroleum products delivered to large industrial customers are assumed to escalate at 6½% per year during this period. This higher rate, compared to the general inflation rate, is probable because of:

● The decreasing domestic production of natural gas and crude oil;
● Increasing demand for these products;
● Prices of imported gas, crude oil, and refined products will follow the inflation rate of the free world rather than that of the U. S.; and
● The high cost of synthetic products manufactured from coal and/or oil shale.

TABLE 8.

ASSUMED U. S. INFLATION RATES

	Inflation Rate %/year 1980 to 2000
Natural gas	6.5
Petroleum products and crude oil	6.5
Coal delivered to large customers	5.0
General goods and services	5.0
Electricity	5.0
Industrial products	5.0
Industrial construction costs	5.0

TABLE 9.

MOST LIKELY ENERGY PRICES, U. S., 1980-2000

1980 $

	1980	2000
Natural gas, $/MBtu	3.15	4.24
High sulfur resid, $/Bbl	15.00	20.20
High sulfur resid, $/MBtu	2.35	3.17
Coal delivered to large industrial plants, $/MBtu		
East Coast	1.04	1.04
Mid-Continent	0.96	0.96
Gulf Coast	1.54	1.54
West Coast	1.25	1.25

Estimates of future costs for manufacturing refinery hydrogen have been calculated for the period 1980-2000 using 1980 $. The procedure involves comparing escalation rates during this period for each component of total cost with the general escalation rate of 5.0% per year for the general economy. Thus, if natural gas increases at 6.5% per year, the natural gas price for 1980 in 1980 $ is increased by about 1.5% per year to obtain the natural gas price (in 1980 $) for the year in question. For example, assume the natural gas price in 1980 (1980 $) is $3.15 per million Btu. The price in the year 2000 would then be:

$$\$ \ 3.15 \ \times \ (1.065)^{20} = \$11.10 \ \text{in current (2000) } \$$$

$$\frac{11.10}{(1.05)^{20}} = \$4.18 \ \text{in 1980 } \$$$

The prices of natural gas or an alternate light hydrocarbon reformer feed, resid, and electricity are assumed to apply to all geographical locations in the U. S. However, delivered-coal prices vary from one geographical location to another. Mine-mouth coal prices are affected by mining costs and coal quality. Transportation costs for moving the coal to manufacturing sites have been assumed to be proportional to the transportation distance.

Coal prices delivered to manufacturing sites on the East Coast, the Mid-Continent (Illinois), the Gulf Coast, and the West Coast were estimated and the transportation cost in unit trains was added to the estimated mine-mouth coal prices to provide estimated prices of coal delivered to the manufacturing sites as are shown in Table 9.

High sulfur coal would be the optimum feed stock for coal gasification plants since the sulfur content of the feed stock has little effect on manufacturing costs and high sulfur coal probably will be priced substantially below low sulfur coals of equivalent heating value.

Investments and manufacturing costs for refinery hydrogen for both 1980 and the year 2000 for plants of 100 million SCF/D capacity located in the Mid-Continent area are summarized in Table 10 and Figure 7. These data compare steam reforming, resid partial oxidation, K-T coal gasification, and new coal gasification. These data include investment data for any year and the operating costs for both 1980 and the year 2000; all data being expressed in 1980 $.

TABLE 10.

ECONOMICS OF MANUFACTURING REFINERY HYDROGEN
100 MILLION SCF/D PLANTS, MID-CONTINENT LOCATION

1980 $

	Natural Gas Steam Reforming	Residuum Partial Oxidation	Coal Gasification K-T	Coal Gasification New
Investment, $ millions, 1980 or 2000	63	159	246	200
Cost of feed stock, $/MBtu				
1980	3.15	2.35	0.96	0.96
2000	4.24	3.17	0.96	0.96
Manufacturing costs, 1980, $MBtu H_2 product				
Feed stock	4.09	3.99	1.61	1.49
All other production costs	0.92	2.30	3.56	2.92
20% before tax return	1.11	3.00	4.64	3.76
Total	6.12	9.29	9.81	8.17
Total manufacturing costs for year 2000, $/MBtu H_2 product	7.54	10.68	9.81	8.17

Low investment cost is the outstanding advantage
for steam reforming. Compared to the $63 million (100%)
investment for steam reforming, the investment for
resid partial oxidation is $159 million (252%), for K-T
coal gasification is $246 million (391%), and for new
coal gasification is $200 million (318%).

Table 10 and Figure 7 also summarize the manufacturing cost of hydrogen produced by these processes
in 1980 $ for the years 1980 and 2000. These calculated
costs for this study are identical with those presented
in the 1977 study (Reference 1), except that costs in
this study for resid partial oxidation are somewhat
higher than those in the 1977 study. An error in resid
feed stock requirements was made in the 1977 study and
this has been corrected in this current study. The
total production cost, including 20%/year before tax
return on investment for the year 1980, is $6.12 per
million Btu (MBtu) (100%) for hydrogen produced by
steam reforming. Corresponding costs are $9.29 (152%)
for resid partial oxidation, $9.81 (160%) for K-T coal
gasification, and $8.17 (134%)for new coal gasification.

The basic conclusion from these economic
comparisons is that steam reforming will continue to be
the preferred method of producing refinery hydrogen
throughout the 1980-2000 period. This advantage for
reforming results from both its substantial investment
advantage and its minimum total manufacturing cost
including return.

The data shown in Table 11 and Figure 8 compare
the investment and operating costs for refinery
hydrogen manufacture by the new coal gasification
process for four major refining centers: East Coast,
Mid-Continent, Gulf Coast, and West Coast. This
comparison was made to determine whether the most
favorable coal gasification system might be attractive
at either of the major refining centers.

The investment cost for the new coal gasification
process varies somewhat from one refining center to
another because the quality of the coal available in
these areas is not the same. For example, the most
likely coal available to East Coast refineries is West
Virginia coal of 26 MBtu/ton heating value. The
Illinois coal that is the likely choice for Mid-
Continent refineries has a heating value of 23 MBtu/
ton and the Wyoming coal that would probably be optimum
for Gulf Coast and West Coast refineries has a heating
value of only 16 MBtu/ton.

The cost of coal delivered to these refining
centers is also shown in Table 11. This cost is $0.96
for the Mid-Continent location, $1.04 for the East Coast,
$1.54 for the Gulf Coast, and $1.25 for the West Coast.

TABLE 11.

ECONOMICS OF MANUFACTURING REFINERY HYDROGEN BY NEW COAL PROCESS
AT FOUR REFINING CENTERS

100 million SCF/D plants, 1980 $

	East Coast	Mid-Continent	Gulf Coast	West Coast
Coal source	W. Virginia	Illinois	Wyoming	Wyoming
Investment, $ millions, 1980 or 2000	191	200	224	224
Cost of coal delivered to plant, $/MBtu, 1980 or 2000	1.04	0.96	1.54	1.25
Manufacturing costs, $/MBtu H$_2$ product, 1980 or 2000				
Coal feed stock	1.62	1.49	2.41	1.96
All other production costs	2.78	2.92	3.23	3.22
20% before tax return	3.58	3.76	4.21	4.21
TOTAL	7.98	8.17	9.85	9.39

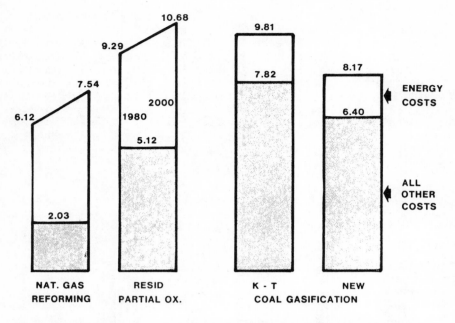

Figure 7. Hydrogen manufacturing costs—midcontinent location (dollars per MBtu–100 Mscf/D plants—1980 dollars)

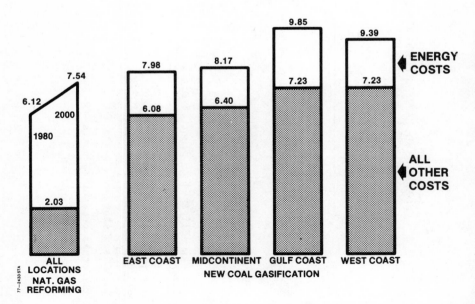

Figure 8. Hydrogen manufacturing costs—new coal gasification (dollars per MBtu–100 Mscf/D plants—1980 dollars)

The manufacturing cost for refinery hydrogen including 20%/year before tax return for each of these locations is $6.12/MBtu for steam reforming in the year 1980 and $7.54 for the year 2000. The cost of producing refinery hydrogen by new coal gasification is somewhat higher than reforming at East Coast and Mid-Continent refining centers and substantially higher than reforming at Gulf Coast and West Coast centers. The conclusion from this comparison is that the new coal gasification process will be more expensive than steam reforming at each of these four major refining centers throughout the 1980-2000 period.

It is apparent that the coal and hydrocarbon prices assumed in this study may not apply to specific refineries. It is possible that long-term coal supplies could be made available to certain refineries at prices less than those assumed in this study. In these locations, new coal gasification might be an attractive means of producing refinery hydrogen. Such a possibility might occur at Gulf Coast refining centers. If East Texas lignite could be delivered to the Gulf Coast refineries at substantially less than the $1.54/MBtu figure used in this study, coal gasification might be an attractive method of producing refinery hydrogen.

Abstract

Hydrogen, a key processing agent in petroleum-refining operations, is consumed in a variety of hydro-desulfurization and hydrocracking operations. Some of the required hydrogen is produced as a by-product of naphtha reforming conducted for octane improvement. The greater portion of the required hydrogen is produced by steam reforming of natural gas (or some alternative light hydrocarbon), although partial oxidation of residuum is used to a limited extent. In the U. S., hydrogen produced for petroleum-refining operations is the second largest industrial use of hydrogen and is about one-half of that required for ammonia manufacture, the largest single industrial use. Requirements for refinery hydrogen will increase gradually during the 1980-2000 period as desulfurization and hydrocracking operations increase relative to the quantity of crude processed. By the year 2000, hydrogen willfully produced for U. S. refining operations is likely to be substantially above the current requirement of about 1.8 billion SCF/D. This willfully-produced hydrogen does not include low pressure, low purity hydrogen produced as a by-product of naphtha reforming for octane improvement.

94   HYDROGEN: PRODUCTION AND MARKETING

Oil and natural gas prices are expected to increase
relative to those for coal during the 1980-2000 period.
Estimates from this study indicate that steam reforming
of natural gas or light naphtha will continue to be more
attractive than coal gasification for producing refinery
hydrogen. This advantage for steam reforming is due to
the low investment and production costs (including 20%
before tax return on investment) relative to the costs
for coal gasification.

References Cited

1.   Corneil, Hampton G., Heinzelmann, Fred J., and
     Nicholson, Edward W. S., Exxon Research and
     Engineering Company, Linden, N. J., "Production
     Economics for Hydrogen, Ammonia and Methanol During
     the 1980-2000 Period," BNL-50663, April, 1977.

2.   Brown, E. C., Eccles, R. M., Lukk, G. G., and
     Rabolini, F. R., Exxon Research and Engineering
     Company, Florham Park, N. J., "Residuum Processing
     for Conversion," AM-76-38, paper presented at 1976
     NPRA Annual Meeting, March 30, 1976, San Antonio,
     Texas.

3.   Kelly, James H., and Laumann, Eugene A., Jet
     Propulsion Laboratory, Pasadena, California,
     "Hydrogen Tomorrow, Demands and Technology
     Requirements," December, 1976, JPL 5040-1, prepared
     under contract NAS 7-100.

4.   Gregory, Derek P., Pangborn, Jon B., and Gillis,
     Jay C., Institute of Gas Technology, 3424 South
     State Street, Chicago, Illinois, 60616, "Survey of
     Hydrogen Production and Utilization Methods," Vol. 1,
     Executive Summary; Vol. 2, Discussion; Vol. 3,
     Appendices, August, 1975.

RECEIVED July 12, 1979.

Hydrogen Production from Partial Oxidation of Residual Fuel Oil

C. L. REED and C. J. KUHRE

Process Engineering–Refining Department, Shell Oil Company,
1100 Milam Building, P.O. Box 3105, Houston, TX 77001

In the past several years, forecasting the future availability and prices of hydrocarbon feedstocks for hydrogen
manufacture has been a very difficult problem. Today, in
addition to the supply and demand forces of the market place,
the indeterminate influence of future governmental regulations
greatly magnifies the dilemma faced by a company trying to
select a long-term economical feedstock for a new hydrogen
plant. For many years steam reforming of natural gas to produce synthesis gas has been a favored basis for hydrogen plant
operation. However, the potential for increased price of
natural gas has now directed attention to selection of heavier
hydrocarbons as feedstock for hydrogen plants.

A process which is well-suited to the basic requirement
of providing hydrogen and carbon monoxide for hydrogen
manufacture is the partial oxidation of hydrocarbons; i.e.,
the combustion of hydrocarbons with a limited amount of oxygen
to yield hydrogen and carbon monoxide. The process is also
referred to as gasification because it converts liquid hydrocarbons to gaseous products, essentially H_2 and CO with only
small amounts of CO_2 and CH_4 when high-purity oxygen is used
as the oxidant. The product gas, often called synthesis gas
or "syngas", can then be converted to high-purity hydrogen as
well as a great variety of other chemical products. The
ratio of hydrogen to carbon monoxide in the synthesis gas
product from the partial oxidation unit is mainly a function
of the hydrogen and carbon content of the feedstock. However,
it is possible to exchange over a CO-shift catalyst the carbon
monoxide of the synthesis gas for hydrogen from steam while
simultaneously producing carbon dioxide.

The processing sequence chosen for this paper, shown in
Figure 1, is a conventional hydrogen plant flow scheme with
each process section having been commercially well-proven.
The synthesis gas product from the Shell Gasification Process
(SGP) unit is treated for sulfur removal in a Shell Sulfinol
unit. The CO-Shift unit includes high-temperature shift cata-

0-8412-0522-1/80/47-116-095$06.75
© 1980 American Chemical Society

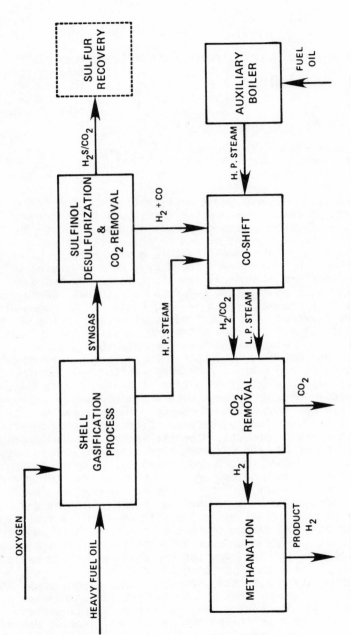

Figure 1. Processing scheme for hydrogen manufacture

lyst followed by low-temperature shift catalyst. The heat
released during the exothermic CO-shift reaction is efficiently
recovered by a complex heat exchange scheme which raises
high-pressure superheated steam and low-pressure steam in
addition to supplying reheat for several intermediate gas
streams. Absorption in hot potassium carbonate was arbitrarily
selected for CO_2 removal from among several very acceptable
commercially well-proven processes; e.g., Sulfinol, Rectisol,
Selexol, or Amine Guard. A Methanation unit is used to convert
the small quantity of residual carbon oxides to methane;
thus producing high purity (> 98% vol H_2) hydrogen.
     The heat and material balances for an SGP-based hydrogen
plant along with the associated costs have been developed for
a plant sized to produce 100 X 10^6 scf/sd of hydrogen. The
hydrocarbon material balance is contained in Tables I, IIA
and IIB. A summary of the estimated capital investment,
identified by plant section, is given in Table III. Table IV
contains a summary of operating cost calculations. Additional
economics details are given in Table V.

The Shell Gasification Process

     The Shell Gasification Process (SGP) is a commercially
proven, highly reliable method for the non-catalytic partial
oxidation of fluid hydrocarbons. The process has great flexibility with respect to suitable feedstocks. This is in
marked contrast to steam reforming of natural gas or naphtha
for hydrogen production. Steam reforming is very restrictive
with respect to acceptable feedstock properties, such as
distillation end-point and maximum sulfur content. With the
SGP, it is only necessary that the feedstock be sufficiently
low in viscosity at preheat temperatures to atomize effectively.
Thus, nearly any fluid hydrocarbon can serve as a suitable
feedstock for the SGP provided that ash content is not exces-
sively high. The following hydrocarbon streams have been
used as SGP feedstocks: natural gas, refinery gases, naphtha,
light fuel oil, heavy fuel oil, heavy residuals (including
asphalt), and whole crude oil. Heavy fuel oil and refinery
residual oil streams are the predominant feedstocks. These
materials are processed in about 70% of the operating SGP
units. Some SGP units have been designed to switch between
gaseous and liquid feedstocks with only a short shutdown
required to switch feed systems. Of course, the downstream
equipment must be designed to handle the change in gas composition resulting from such a wide change in feedstock type.
     The SGP crude synthesis gas is utilized to manufacture
a wide range of end-products. Ammonia is the predominant
product with over 50% of the SGP capacity being devoted to
manufacturing this material. Methanol accounts for the second
largest usage of SGP synthesis gas with oxo-products in third

TABLE I.

GASIFICATION OF HEAVY FUEL OIL
Shell Gasification Process

Basis: Production of 100 X 10^6 scf/stream day H_2

SGP Feedstock Type	Heavy Fuel Oil
SGP Feedstock Properties	
Gravity, °API	14.2
Specific Gravity, 60°F/60°F	0.97
C/H Ratio, weight	7.49
H/C Ratio, atomic	1.59
Sulfur, %wt	3.50
Ash, %wt	0.07
SGP Feedstock, lb/sh (150°F)	88,050
SGP Oxygen Feed,[a] lb/sh (100°F)	94,070
SGP Naphtha Net Usage For Carbon Recovery,[b] lb/sh	440
Fuel Oil For Auxiliary Boiler, lb/sh	9,270

[a] Expressed as 100% O_2; actual O_2 purity 99.50 % vol. Pressure is 920 psig.

[b] Constitutes feed to gasification reactor over and above the heavy fuel oil feedstock. Soot production is recycled 100% to gasification reactor.

TABLE IIA.

SHELL GASIFICATION PROCESS AND SULFINOL
PRODUCT GAS COMPOSITIONS AND RATES

Basis: Production of 100×10^6 scf/stream day H_2
Feedstock to the SGP is heavy fuel oil
- see Table 1 for feedstock and oxygen
requirements.

	SGP Scrubber Top[a]	Sulfinol Desulfurized Gas[b]	Acid Gas[c]
Composition, %vol (Dry)			
Hydrogen	46.72	48.98	3.11
Carbon Monoxide	48.16	50.46	3.65
Carbon Dioxide	3.80	< 0.01	77.10
Methane	0.30	0.31	0.05
Nitrogen + Argon	0.23	0.24	0.02
Hydrogen Sulfide	0.76	< 5 ppmv	16.07
Carbonyl Sulfide	0.03		–
Flow Rate			
lb-moles/sh	11,917	11,329	592
10^6 scf/sd	108.5	103.2	5.4
Pressure, psig	775	765	12

a) Product gas from the Shell Gasification Process which is feed gas to the Sulfinol unit.

b) Feed gas to CO-Shift unit.

c) Acid-gas from the Sulfinol unit must be processed in a Sulfur Recovery unit. Costs for the Sulfur Recovery unit are not included in the costs presented in this paper.

TABLE IIB.

CO-SHIFT UNIT, CO_2-REMOVAL UNIT AND METHANATION UNIT PRODUCT GAS COMPOSITIONS AND RATES

Basis: Production of 100 X 10^6 scf/stream day H_2. Feedstock to the SGP is heavy fuel oil

	CO-SHIFT PRODUCT GAS[a]	CO_2-REMOVAL TREATED GAS[b]	METHANATION PRODUCT GAS[c]
Composition, %vol (Dry)			
Hydrogen	65.96	98.84	98.82
Carbon Monoxide	0.40	0.60	—
Carbon Dioxide	33.27	< 0.01	—
Methane	0.21	0.31	0.94
Nitrogen + Argon	0.16	0.24	0.24
Flow Rate			
lb-moles/sh	16,979	11,329	11,123
10^6 scf/sd	154.6	103.2	101.3[d]
Pressure, psig	675	655	630

a) Product gas from the CO-Shift unit is feed gas to the CO_2-Removal unit.

b) Treated gas from the CO_2-Removal unit is feed gas to the Methanation unit.

c) Product gas from the Methanation unit is the plant hydrogen product.

d) Pure hydrogen in plant product gas is 100.1 X 10^6 scf/sd.

TABLE III.

SGP BASED HYDROGEN PLANT

ESTIMATED CAPITAL INVESTMENT

Basis: Production of 100 X 10^6 scf/stream day H_2.
Feedstock to the SGP is heavy fuel oil

Plant Section	Estimated Installed Capital Cost,[a]10^6
Gasification	27.0
Sulfinol	5.4
CO-Shift	14.0
Auxiliary Oil-Fired Boiler	7.2
CO_2-Removal & Methanation	9.4
Offsites (35%)[b]	22.0
Total	85.0

[a]Basis: mid-1979, U.S. Gulf Coast, installed costs.
No allowance made for interest during construction.

[b]Offsites estimated at 35% of onsite capital to cover such items as
cooling water, feed water preparation, utilities, feedstock storage,
effluent treating, and other minor offsites required for a plant
installed adjacent to an existing facility. Oxygen is considered to
be delivered "over the fence" at the required pressure and no capital
allowances are included for this supply.

TABLE IV.

SGP BASED HYDROGEN PLANT

ESTIMATED OPERATING COSTS

100 X 10^6 scf/stream day H_2
Available at 630 psig
330a) operating days per year

Variable Costs	Consumed Per Year	$ Per Unit	Cost b) $ 10^6 Per Year	$ Per 10^3 scf H_2
SGP Feedstock	2.053 X 10^6 bbl	15.17/bbl e)	31.14	0.94
SGP Naphtha	14 X 10^3 bbl	18.00/bbl f)	0.25	0.01
SGP Oxygen	373 X 10^3 s tons	32.00/s ton	11.94	0.36
Boiler Fuel	0.245 X 10^6 bbl	17.39/bbl g)	4.26	0.13
Chemicals & Catalyst	-	-	0.22	0.01
Electric Power	65.74 X 10^6 kwh	0.03/kwh	1.97	0.06
Total Variable Costs			49.78	1.51

Fixed Costs

Operating Labor [c]	0.86	0.03
Maintenance [c]	4.52	0.14
Administrative	1.70	0.05
Insurance	1.70	0.05
Capital Charge [d]	14.45	0.44
Total Fixed Costs	23.23	0.71
Total Fixed plus Variable Costs	73.01	2.22

a) Specified by Symposium Co-Chairman. (See the Table 5 for additional economics details.) Well maintained SGP and Sulfinol units normally operate in excess of 347 days per year (>95% onstream factor) depending upon reliability of oxygen supply.
b) Basis mid-1979 costs.
c) Includes direct labor (6 operating jobs per shift), direct supervision, support labor, and payroll extras-fringe benefits, etc.
d) Annual charge of 17% of total installed capital.
e) Arbitrarily assumed for mid-1979 at $2.40/MM Btu GHV.
f) Arbitrarily assumed for mid-1979.
g) Arbitrarily assumed for mid-1979 at $3.00/MM Btu GHV.

TABLE V.
ECONOMIC BASIS

	Basis
Time Frame	Mid-1979
Project Life	20 years[1])
Operating Factor	330 days/year
Cost of capital	10%[1])
Annual Operating Cost	
Utilities	
Power ($/kwh)	0.03
Water ($/$10^3$ gal)	0.05
Fuel Oil Feed to SGP ($/$10^6$ Btu)	2.40
Fuel Oil for Auxiliary Boiler ($/$10^6$ Btu)	3.00
Operations	
Labor	$18,000/man-year
Supervision	15% of operating labor
Materials	5% of operating labor

Maintenance

    Labor   2% of facilities investment

    Supervision   15% of maintenance labor

    Materials   2% of facilities investment

Administrative and support labor   20% of all operating + maintenance labor

Payroll extras - fringe benefits, etc.   20% of all operating + maintenance labor

Insurance   2% of facilities investment

General administrative expenses   2% of facilities investment

Taxes (Local, State & Federal)   50% of gross profit[1]
(No investment tax credit)

Depreciation   Straight line[1])

[1])An annual capital charge of 17% of total capital investment is about equal to a DCF rate of 10% when the above data are used in the DCF calculation.

place. Hydrogen and reducing gas are also produced. Town gas
was produced for several years by using hot carburation with
naphtha to increase the heating value of the synthesis gas to
a level of about 500 Btu/scf (GHV). Medium-Btu fuel gas of
about 330 Btu/scf (GHV) can be produced by partial oxidation
with oxygen without carburation, and low-Btu fuel gas of about
120-150 Btu/scf (GHV) can be obtained by using air as the oxidant.

Well over 110 SGP reactors at over 35 different plant sites
have been placed in operation around the world since the process
was first commercialized in 1956. Several SGP units were
started up in 1977-78, including two plants located in the
U.S.A.

SGP Chemistry

The process is carried out by injecting preheated hydrocarbon, preheated oxygen, and steam through a specially
designed burner into a closed combustion vessel, where partial
oxidation occurs in the range of 2350°F to 2550°F (1290°C -
1400°C). "Partial Oxidation" describes the net effect of a
number of component reactions that occur with hydrocarbons
within a reactor supplied with less than stoichiometric
oxygen for complete combustion. The overall reaction is
represented by:

$$C_nH_m + (n/2)\ O_2 \longrightarrow n\ CO + (m/2)\ H_2$$

The overall process can be divided into three phases as
described below:

A. Heating-up and Cracking Phase. In the fuel injection
region of the reactor, preheated hydrocarbons leaving
the atomizer are intimately contacted with the steam/
preheated oxygen mixture. The atomized hydrocarbon is
heated and vaporized by back radiation from the flame
front and the reactor walls. Some cracking to carbon,
methane and hydrocarbon radicals occurs during this
brief phase.

B. Reaction Phase. As soon as the fuel and oxidant
reach the ignition temperature, hydrocarbons react
with oxygen according to the highly exothermic combustion
reaction. Practically all of the available oxygen is
consumed in this phase.

$$C_nH_m + (n + m/4)\ O_2 \longrightarrow n\ CO_2 + (m/2)\ H_2O$$

The remaining hydrocarbons, which have not been oxidized,
react endothermically with steam and the combustion pro-

ducts from the primary reaction. The main endothermic
reaction is the reforming of hydrocarbon by water vapor:

$$C_nH_m + n\ H_2O \rightleftharpoons n\ CO + (n + m/2)\ H_2$$

Excessive local temperatures can damage the refractory;
thus, it is essential that all reactants be well mixed
so that the endothermic reactions tend to locally balance
the exothermic reactions. In this way, the complex of
reactions is quickly brought to thermal equilibrium,
resulting in a measured temperature within the range of
2350°F to 2550°F.

C.   Soaking Phase. The soaking phase takes place in the
rest of the reactor, where the gas is at high temperature.
Minor changes in gas composition occur due to secondary
reactions of methane and carbon. As the reaction rates
are relatively low, the methane content is higher than
would be expected from equilibrium. During the soaking
phase, a portion of the carbon also disappears by reactions
with CO_2 and steam. However, some carbon is always
present in the product gas from the reactor in a quantity
equivalent to about 1-3% wt of the oil feed. Natural gas
feedstock produces only a very small amount of residual
carbon; i.e., about 0.02% wt of the gas feedstock.

The final composition of the SGP reactor product gas is
established by the water-gas shift equilibrium at the
reactor outlet/waste-heat exchanger inlet where rapid
cooling begins.

$$CO + H_2O \rightleftharpoons CO_2 + H_2$$

In the strongly reducing atmosphere of the SGP reactor,
sulfur compounds in the hydrocarbon feedstock react to
form hydrogen sulfide (H_2S) and small amounts of carbonyl
sulfide (COS) in a molar ratio of approximately 24:1.
Sulfur content of the feedstock presents no serious problem
for SGP; high-sulfur (e.g., 5%wt S) petroleum residues
can serve as SGP feedstocks. The H_2S and COS can be
readily removed from the SGP product gas by any one of
several well-proven desulfurization processes such as the
Shell Sulfinol Process.

SGP Process Description

The Shell Gasification Process consists essentially of:
1) the gasification reactor, 2) the waste-heat exchanger for
heat recovery from the hot reactor gas, 3) the economizer heat
exchanger for further heat recovery, 4) the carbon removal

system for separating carbon from the reactor product gas, and
5) the carbon recovery system for recycle of carbon. Figure 2
is a simplified flow diagram for the SGP as applied to gasification of an oil feedstock.

The gasification reactor is a vertical, steel pressure
vessel with a refractory lining. There are no internal baffles
or catalyst beds. Preheated hydrocarbon feedstock and oxidant
are fed under precise flow control to the specially designed
combustor in the top of the reactor. Steam is pre-mixed
with the oxygen to serve as a flame moderator. For liquid
feedstocks, the oxidant enters the reactor as a rotating vortex
around the hydrocarbon vortex spray in the combustion zone.
This promotes effective mixing of hydrocarbon and oxidant.
The multiple-layer refractory lining is designed to withstand
the high temperature in the reactor. Commercially available
refractory materials are used; they have given very good service life. Adequate residence time is provided in the reactor
to permit the partial oxidation reactions to approach equilibrium. Reactor pressures up to 835 psig have been commercially
demonstrated; gasification at higher pressure is possible.
SGP reactors have also been designed to operate at a pressure
as low as about 15 psig. The reactor pressure is usually
set by downstream processing requirements. For example, the
pressure required in a low-pressure methanol unit can be obtained at the outlet of the SGP without compression of the
synthesis gas.

An important feature of the SGP is the capability
for recovering much of the heat in the reactor product gas
by generating high-pressure steam. The hot gas from the
reactor flows directly to a waste-heat exchanger of special
design where it passes through helical coils mounted in the
exchanger shell. The sum of the sensible heat recovered from
the reactor effluent gas (by raising steam) plus the potential
heat of combustion represented by the product gas itself is
equal to about 95% of the hydrocarbon feedstock heating value
(GHV). Soot in the hot gas from the partial oxidation
reactor could present serious deposition problems in conventional exchanger tubes. The use of helical tubes and proper
gas velocity gives very long service life without the need
for periodic cleaning and without impairment of heat transfer.
The steam is generated at a pressure at least 150 psi greater
than the reactor pressure so that it can be used directly as
moderating steam. Waste-heat exchangers for the SGP have been
designed for steam pressures to about 1,500 psig.

The gas leaves the waste-heat exchanger at a temperature
somewhat greater than the generated steam temperature. Additional heat is recovered from the gas in an economizer by
heat exchange with feedwater for the waste-heat exchanger.
The split of heat recovery duties between the waste-heat
exchanger and the economizer is optimized during the design

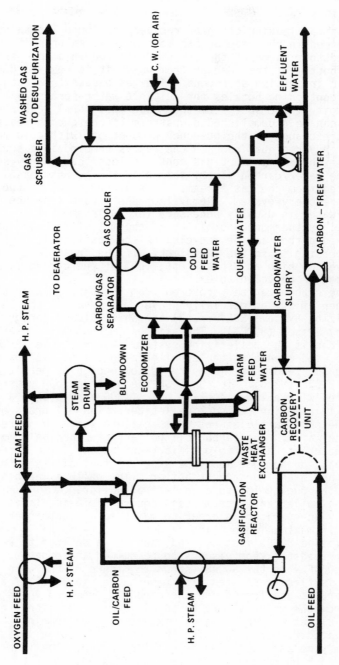

Figure 2. Shell gasification process

phase of an SGP project.

The gas leaving the heat recovery equipment contains the soot formed in the reactor and most of any ash present in the feedstock; some of the ash is deposited in a void space in the bottom of the reactor from which it is removed during periodic inspection shutdowns. The gas passes to a quench vessel containing nozzles for multiple water-sprays which scrub most of the soot from the gas. Additional heat recovery can be accomplished downstream of the quench vessel by heat exchange of the quenched product gas with cold feed water. Any residual soot in the gas is removed in a scrubber column. The SGP product gas contains less than 5 ppmv soot.

The water/carbon slurry formed in the quench vessel is separated from the gas stream. This slurry flows to the carbon recovery system. Removal of the carbon from the slurry water is usually necessary for environmental considerations. Furthermore, for improved thermal efficiency, the recovered carbon can be recycled to the gasification reactor by dispersing it in the feedstock. If the fresh feed does not have too high an ash content, 100% of the carbon formed can often be recycled to extinction.

The carbon can be recovered from the water/carbon slurry by either of two commercially proven methods: 1) the Shell Pelletizing System (SPS) or, 2) the Shell Closed Carbon Recovery System (SCCRS). The simplest method of recovering the carbon is the pelletizing system. However, the SPS can only be used when a suitable pelletizing oil is available and the SGP feedstock viscosity is low enough to permit pumping at 200°F (93°C). In the SPS, the water/carbon slurry is contacted with a low-viscosity oil, (<300 centistokes at 200°F) in a device called a pelletizer. The oil preferentially wets the soot particles and forms pellets which are screened from the water. The carbon/oil pellets can then be homogenized into the oil feed to the gasification reactor, used directly as a fuel, or homogenized in a separate fuel oil to be burned in another plant furnace or boiler. The pellets can also be mixed with fuel going to a coal-fired boiler.

The pelletizer system does require periodic visual surveillance by an operator in order to maintain a smooth operation. Hence, if maximum automation is desired, the SCCRS is recommended. The SCCRS must be used when it is desired to recycle soot in feedstock too viscous to be pumped at temperatures below 200°F. In the SCCRS (shown in Figure 3) the water/carbon slurry is first contacted with naphtha so that carbon/naphtha agglomerates are formed and the carbon is removed in this form from the water slurry to be mixed with additional naphtha. The resultant carbon/ naphtha mixture is then combined, at an elevated temperature, with the hot gasification feedstock which may be as viscous as deasphalter pitch. The feedstock/ carbon/naphtha mixture is heated and flashed, and then fed to a naphtha stripper where naphtha is recovered for recycle to the carbon/water separation step. The carbon remains dispersed in the hot feedstock leaving the bottom of the naphtha stripper column and is recycled to the gasification reactor.

Desulfurization - Sulfinol Process Description

The Shell Sulfinol Process is used for removal of acidic constituents such as H_2S, CO_2, COS, etc. from a gas stream. Improved performance over other processes is due to the use of an organic solvent, Sulfolane (tetrahydrothiophene dioxide), mixed with an aqueous alkanolamine. Relative proportions of Sulfolane, alkanolamine, and water, as well as the operating conditions, are tailored for each specific application. Simultaneous physical and chemical absorption under feed gas conditions is provided by this Sulfinol solvent. Regeneration is accomplished by release of the acidic constituents at near atmospheric pressure and a somewhat elevated temperature. The flow scheme (Figure 4) is very similar to that of an aqueous alkanolamine system since it involves only absorption, regeneration and heat exchange under typical alkanolamine treater conditions.

The SGP product gas is contacted countercurrently in an absorber column by the Sulfinol solvent which absorbs the acidic components from the SGP gas. The rich solvent then flows through a heat exchanger where it is heated by the hot, lean solvent from the bottom of the regenerator. The hot, rich solvent is regenerated by pressure reduction, further heating, and stripping in the regenerator column where the acid gases are liberated. The lean solvent is cooled by heat exchange with the rich solvent before returning via a cooler to the absorber where it begins its scrubbing cycle again.

The advantages which can be obtained when using the Sulfinol Process as compared to other treating systems are: 1) reduced solvent circulation due to higher acid gas solubilities; 2) a lower energy requirement for solvent regeneration; 3) a treated gas meeting the specifications required for downstream processing; and 4) an acid-gas product suitable for blending into other feed streams to a Claus sulfur recovery unit; 5) improved operability and smaller equipment, resulting from the non-foaming characteristics of the solvent; 6) low corrosion rates; and 7) high on-stream factor. The Sulfinol Process also has the attractive feature in that it can achieve nearly complete removal of carbonyl sulfide (COS). Well over 100 Sulfinol installations are in operation throughout the world. Most of these units are treating natural gas; however, more than 10 Sulfinol units are treating product gas from partial oxidation plants.

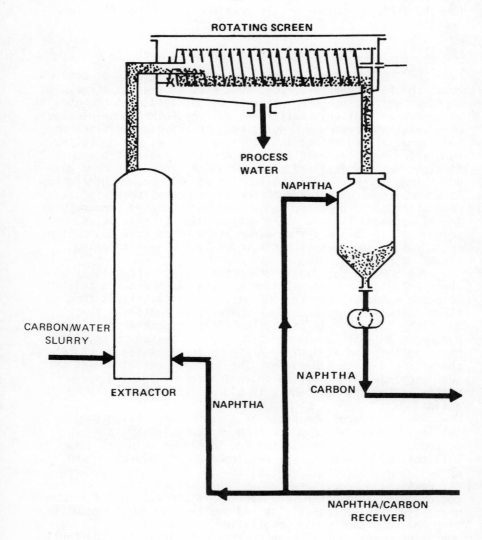

Figure 3.   Shell closed

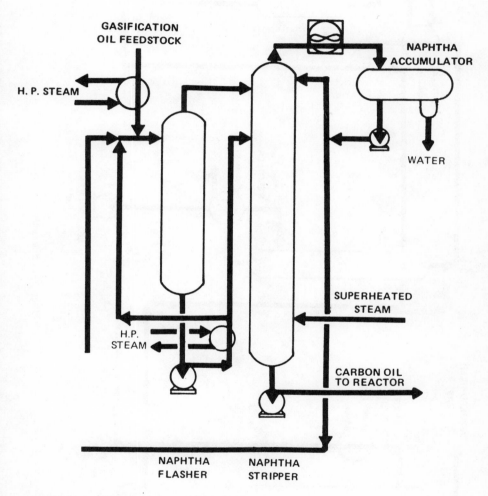

carbon recovery system

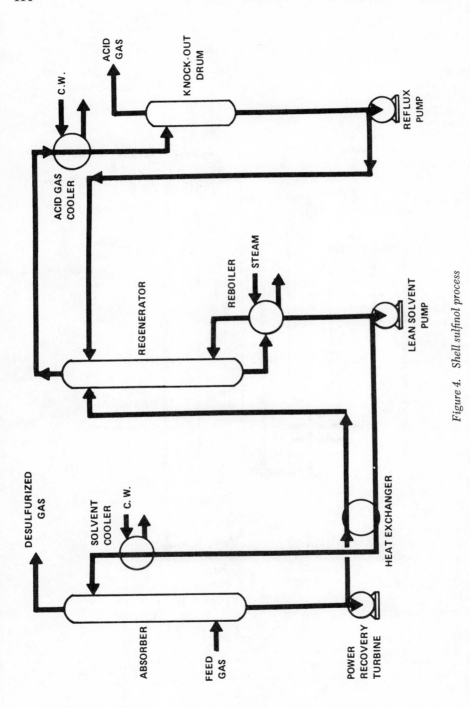

Figure 4. Shell sulfinol process

Sulfur Recovery

The acid gas from the Sulfinol regenerator must be disposed of in an environmentally acceptable manner. The Claus process offers an effective means for converting nearly all of the sulfur in the acid gas to saleable elemental sulfur. The tail gas from the Claus plant still contains some sulfur compounds. To minimize sulfur emissions from the plant, the Claus tail gas can be fed to a Shell Claus Off-gas Treating (SCOT) unit where most of this sulfur is recovered and recycled to the Claus plant. With use of the SCOT Process, additional marketable sulfur is recovered within the Claus plant while tail gas sulfur emissions are substantially reduced, to typically less than 250 ppmv.

The SCOT Process has been accepted by industry as an excellent means of reducing Claus tail gas sulfur emissions. There are 36 SCOT units in the USA: 18 in operation, and 18 others in various stages of design or construction. In the rest of the world, there are 25 units: 18 in operation, and 7 others in various stages of design or construction. World-wide, these 61 units, either in operation or firmly committed for installation, represent an aggregate Claus plant capacity of over 10,000 long tons/day sulfur. Claus plant capacities associated with these SCOT units range from 5 to 2,100 long tons/day sulfur.

When a partial oxidation synthesis gas is treated for H_2S and CO_2 removal, the resultant acid gas from the treating system is usually quite lean in H_2S concentration due to the relatively large amount of CO_2. For example, see the acid gas composition in Table IIA where H_2S is only 16.07% vol of the Sulfinol acid gas (dry basis). This H_2S concentration is near the minimum level which would permit processing the acid-gas in a Claus unit. Moreover, a special splitflow type Claus unit or other unusual procedures would be required to handle this low concentration.

Shell also offers for license a H_2S-selective version of the Shell ADIP Process. The ADIP process, which has a flow scheme very similar to Sulfinol, can be used to treat the Sulfinol acid gas to raise the H_2S concentration by selectively rejecting the CO_2. Some integration of the SCOT process with the ADIP process is often possible; thus, reducing overall equipment and operating costs. Costs for the Claus plant are substantially reduced when "selective" ADIP is applied. Two selective ADIP plants are scheduled to come on stream in the first half of 1979.

A detailed discussion of optimization of sulfur recovery facilities would comprise another paper in itself, and therefore, is outside the scope of this paper. Costs of hydrogen production, which are presented in Figure 5 and Tables III and IV do not include sulfur recovery (Claus, etc.) costs or sulfur

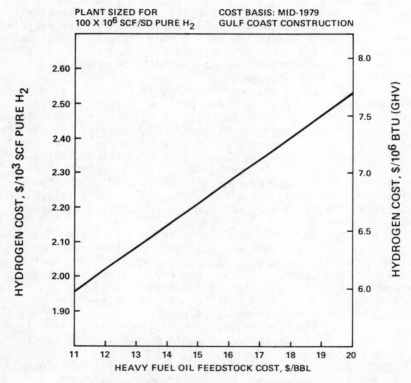

NOTE: PRODUCTION COST INCLUDES A 17% ANNUAL CAPITAL CHARGE.
        OXYGEN IS PURCHASED OVER-THE-FENCE AT PRESSURE FOR
        $32/S TON.

Figure 5.   Cost of hydrogen produced by partial oxidation of heavy fuel oil

product credits.

CO Shift Process Description

The CO-Shift unit reduces the CO content of the Sulfinol treated gas to 0.4% vol on dry product gas. The high-temperature shift (HTS) section comprises three reactors in series with intermediate gas cooling by heat exchange with feed gas and by steam production. Steam feed for the CO-Shift unit is produced in: 1) the SGP waste-heat exchangers, 2) heat recovery boilers within the CO-Shift unit and, 3) an auxiliary boiler. Steam feed to the first HTS reactor is set to give a steam-to-dry gas ratio of 1.5 mol/mol. This is a conservative (high) ratio. If further optimization were carried out for a specific plant location it could indicate that the steam-to-dry gas ratio may be reduced to as low as 1.2 mol/mol with some resultant loss in hydrogen purity and yield, but with a savings in auxiliary boiler fuel and capital costs.

The third HTS reactor will deliver product gas containing about 2.8% vol CO (on dry gas) to the low-temperature shift (LTS) section. Heat recovered from the LTS effluent is used to raise 90 psig steam and to preheat boiler feed water. Zinc oxide beds are used to safeguard the sulfur-sensitive LTS catalyst against sulfur poisoning from the expected 1 to 2 ppm H_2S + COS slippage on the Sulfinol unit.

CO_2 Removal and Methanation Process Description

For CO_2 removal, a typical process was included for illustrative purposes, wherein a solvent absorbs the CO_2 and is then regenerated by applying heat. No attempt was made to compare the many possible process choices for CO_2 removal; e.g. another alternative method for CO_2 removal could be the use of Pressure Swing Adsorption units which would produce higher purity hydrogen and require no methanator, but at the cost of a reduced level of hydrogen recovery. The heat required for CO_2 removal solvent regeneration is provided by 90 psig steam raised in the CO-Shift unit. The CO_2-Removal unit is designed to reduce the CO_2 content of the treated gas to less than 50 ppm vol.

The Methanation unit consists of a catalyst bed contained in a reactor vessel, feed-to-product heat exchange, product cooling and a knock-out vessel for separating condensed water. Treated gas from the CO_2-Removal unit is also preheated by heat exchange in the CO-Shift unit to conserve energy.

The principal methanation reactions, given below, illustrate why the purity and yield of hydrogen is influenced by the effectiveness of the CO-Shift and CO_2-Removal units.

$$CO_2 + 4H_2 \rightleftharpoons CH_4 + 2H_2O$$
$$CO + 3H_2 \rightleftharpoons CH_4 + H_2O$$

Since CO-Shift unit product gas contains only 0.40% vol CO and the CO_2-Removal unit treated gas contains only 50 ppm vol CO_2, the production of methane impurity and consumption of hydrogen in the Methanation unit is minimal. Moreover, the methane content of the SGP product gas (0.30% vol CH_4) is significantly lower than the 1 to 3% vol of methane which is normal slippage for a steam-methane reformer (SMR). Hence, a higher purity hydrogen stream is feasible with an SGP-based hydrogen plant versus an SMR-based plant. Alternatively, the residual CO and CO_2 specifications can be relaxed in the CO-Shift and CO_2-Removal units for a SGP-based plant while producing the same purity hydrogen as from an SMR-based plant. Of course, use of Pressure Swing Adsorption for CO_2 removal would produce equivalent ultra-high purity hydrogen from either SGP or SMR-based plants.

Process Economics

Estimated installed capital costs are shown in Table III and estimated operating costs are given in Figure 5 and Table IV for the production of hydrogen from heavy fuel oil by use of the SGP and Sulfinol units along with CO-Shift, CO_2-Removal and Methanation units. The design bases for the process economics case are described below.

Design Bases For SGP-Based Hydrogen Plant

1.   Production of 100×10^6 scf/sd of pure hydrogen. Hydrogen purity is greater than 98% vol.

2.   Heavy (14.2° API), high-sulfur (3.5% wt sulfur) fuel oil feedstock (see Table I).

3.   Operation of the SGP reactors at 835 psig; this leads to a Methanation unit product gas at 630 psig.

4.   Recycle of recovered carbon to extinction using the Shell Closed Carbon Recovery System (Figure 3).

5.   Oxygen is purchased over-the-fence at 920 psig with no energy integration between the air separation plant and the SGP.

6.   Operation for 330 days per year (90% stream
      factor for conventional economic analysis).
      However it should be emphasized that higher
      stream factors (>95%) are achievable depend-
      ing upon reliability of oxygen supply.

For the capacity of the proposed plant in this study,
we have chosen to use two SGP reactor trains feeding into
one scrubber. The units downstream of the SGP would be
single train.

The feedstock requirements for the hydrogen plant as
defined are: 6,221 bbl/sd of heavy fuel oil (Bunker C or
ASTM No. 6), 41 bbl/sd of naphtha, 1,129 short tons/sd of
oxygen at 920 psig, and 744 bbl/sd of ASTM No. 2 fuel oil
for the auxiliary boiler. Oxygen purity is assumed to be
99.5% vol. Lower oxygen purity is acceptable, but there
will then be additional dilution of the product hydrogen by
nitrogen and argon. The small amount of naphtha required is
makeup for the carbon recovery system. This naphtha is ultimately consumed as SGP feedstock. Low-sulfur ASTM No. 6
fuel oil could also be used as the fuel for the auxiliary
boiler instead of ASTM No. 2 fuel oil. High-sulfur fuel oil
should also be considered if the cost of flue gas desulfurization is justified by the lower cost of the fuel oil.

Capital Costs

The estimated installed capital costs (Table III) were
developed using correlated cost information. The capital
costs include 10% for contingencies. The location is assumed to be on the U.S. Gulf Coast.

As shown in Table III, an allowance equal to 35% of process
unit capital cost was assumed for utilities and general facilities on the basis of an established site location; i.e., no
allowances were included for "greensite" requirements of
office buildings, shops, etc.

The total installed capital cost for the SGP based hydrogen
plant is estimated at $85 million, basis installation in mid-1979
on the U.S. Gulf Coast. No allowance is included for interest
during construction.

Operating Costs

Estimated fixed costs; i.e., operating labor, maintenance,
administration, insurance, and capital charge, are shown in
Table IV. Except for the capital charge, the fixed costs contribute relatively little to the total hydrogen cost. The annual
capital charge used in this analysis is 17% of the total installed plant cost. This is equivalent to a simple before-tax
payout of slightly less than six years.

For convenience in developing and illustrating the estimated variable operating costs of Table IV, the following arbitrary values for feed materials and energy were taken for mid-1979. Additional details of the economic basis are given in Table V.

1. Heavy fuel oil, 3.5% wt. sulfur - $15.17/bbl = $2.40/10^6 Btu (GHV).

2. Naphtha, makeup for the carbon recovery system - $18.00/bbl.

3. Electric power cost - $0.03/kwh.

4. Oxygen purchased over-the-fence at 920 psig- $32/s ton.

5. Fuel for the auxiliary boiler, ASTM No. 2, $17.39/bbl = $3.00/10^6 Btu (GHV).

The above values for materials and energy represent one set of projected costs. A summary of these values is given in Table IV where the cost of hydrogen is shown to be $2.22/10^3 scf of H_2. Other energy cost scenarios can be evaluated using Figure 5 which shows hydrogen manufacturing cost as a function of heavy fuel oil feedstock cost with other unit costs being held constant.

The manufacturing cost for hydrogen is very sensitive to the feedstock cost. At about 43% of the total cost, it is by far the largest single cost component (see Table IV). The maintenance cost of $4.52 X 10^6 per year shown in Table IV is quite conservative (high) because the rule-of-thumb percentage values shown in Table V were applied to the total capital including offsites. Maintenance costs for the offsites, as a percent of capital, are expected to be much less than the onsite process equipment. The $32/s ton cost for 920 psig oxygen was selected based on providing the oxygen supplier with approximately 10% return on capital investment, excluding any credit for sale of nitrogen or argon by-products.

A reduction of steam-to-dry gas ratio in the CO-Shift unit to 1.2:1.0 mol/mol as discussed in the CO-Shift process description section would reduce the hydrogen cost by about $0.07/10^3 scf of H_2.

This study case is designed as a free-standing hydrogen plant. Economies of operation could undoubtedly be obtained if this plant can be integrated with the utility system of an existing complex; e.g. the 150,000 lb/hr of steam produced from the auxiliary boiler could be more efficiently included in the capacity of a larger boiler.

The hydrogen product (including methane impurity) from the Methanation unit has a calculated heating value of 330 Btu/scf

(GHV). Hence, the $2.22/10^3$ scf pure H_2 cost given in Table IV can also be stated as $6.73/10^6$ Btu (GHV). The cost of hydrogen in terms of $/10^6 Btu (GHV) is also shown on Figure 5 as a function of feedstock cost.

Summary

The Shell Gasification Process (SGP) is a highly versatile, well-proven, system for producing synthesis gas (H_2+CO) from virtually any available fluid hydrocarbon feedstock ranging from natural gas to asphalt. This feedstock flexibility feature can be especially valuable in times of hydrocarbon scarcity. This H_2 + CO synthesis gas can be processed for sulfur-removal (Sulfinol), CO-shift (HTS/LTS), CO_2-removal, and methanation to produce a high purity (>98% vol H_2) hydrogen.

The estimated manufacturing cost for hydrogen is $2.22/10^3$ scf H_2 based upon feeding a heavy, high-sulfur fuel oil valued at $15.17/bbl ($2.40/10^6 Btu, GHV). An SGP-based hydrogen plant designed to produce 100×10^6 scf/sd of H_2 is estimated to cost $85 million (mid-1979, U.S. Gulf Coast), including an arbitrary allowance of 35% of onsite capital cost for offsites and 10% for contingencies.

RECEIVED July 12, 1979.

Synthetic Gas Production for Methanol: Current and Future Trends

J. A. CAMPS and D. M. TURNBULL

Davy Powergas Inc., P.O. Drawer 5000, Lakeland, FL 33803

Methanol is the one of most easily made organic compounds and is synthesized from a gaseous mixture of carbon monoxide and hydrogen.

$$CO + 2H_2 \qquad CH_3OH$$

The carbon monoxide/hydrogen mixture is traditionally referred to as "Synthesis Gas" and typically includes small percentage of CO_2, CH_4, nitrogen and other inerts.

Thus, although hydrogen is used in methanol production, it can be taken straight from the steam-hydrocarbon reformer and does not require further purification and treatment as in the case of pure hydrogen production or ammonia production. The economics of methanol production are significantly affected by the thermal integration of the reformer (or other gas generation unit) with the rest of the plant.

Synthesis gas may be prepared from any feedstock containing any ratio of carbon and hydrogen and oxygen and not extreme levels of sulphur and nitrogen. Such a definition covers feedstocks ranging from wood, biomass, coal and heavy fuel oils, to naphtha and natural gas.

Naphtha was the initial feedstock for the ICI low pressure methanol process because it was available to ICI at the time they developed the process. However, it is also the most ideal feedstock from a stoichiometric viewpoint; as evidenced by the following reaction.

$$"CH_2" + H_2O \qquad CO + 2H_2 \qquad CH_3OH$$

Naphtha + Steam   Synthesis Gas   Methanol

Depending upon the stoichiometry of the feedstock source, a stoichiometric synthesis gas is achieved by adjusting the carbon/hydrogen ratio via shift conversion

0-8412-0522-1/80/47-116-123$06.00
© 1980 American Chemical Society

and CO_2 removal or CO_2 addition. Alternatively, within certain limits, the flexibility of the ICI LP methanol process allows a hydrogen rich gas to be economically fed to the synthesis loop. The excess hydrogen is purged from the loop and burned as fuel.

The following is a brief description of synthesis gas generation from the various accepted feedstocks.

Reforming of Natural Gas

By far the most widespread method of producing synthesis gas for methanol production today is via steam reforming of naphtha and lighter hydrocarbons. The process routes for both natural gas steam reforming and naphtha reforming are virtually identical, so we shall consider only natural gas reforming. Since the ICI LP methanol process was first introduced in 1967, the design has been modified to enhance energy recovery as it became more economic to do so due to increasing energy costs. Davy's latest process flowsheet is illustrated in Figure 1. Figures 2 and 3 are photographs of two natural gas based methanol plants built by Davy.

Brief Process Description of Davy's Latest High Efficiency Design. Natural gas is desulfurized over an activated carbon bed or zinc oxide to remove sulfur below 0.2 ppm, suitable for natural gas reforming.

The desulfurized gas is then countercurrently contacted against hot water which heats the feedstock and saturates it with water vapor providing 49% of the process steam requirements. (Over 60% for a plant with CO_2 addition).

The feedstock is then preheated to the reformer inlet conditions, and the balance of the process steam is added. The mixed steam/feedstock mixture is then passed to the tubular reformer where it is reformed to synthesis gas over a nickel catalyst contained in high alloy tubes. This reaction is highly endothermic, and the heat of reaction is supplied by firing natural gas and purge gas in a fire box external to these tubes. The heat is then transferred through the tube wall into the catalyst packed reaction zone.

The reformed gas leaves the furnace at a high temperature where high grade heat is recovered successively to a reformed gas boiler, steam superheater process feedstock heater and boiler, feedwater heater. The reformed gas then passes to the distillation area where low grade heat is efficiently recovered via column reboilers and a demineralized water heater.

Finally, the reformed gas passes to the reformed gas cooler where the gas is cooled from 150°F to 100°F against cooling water.

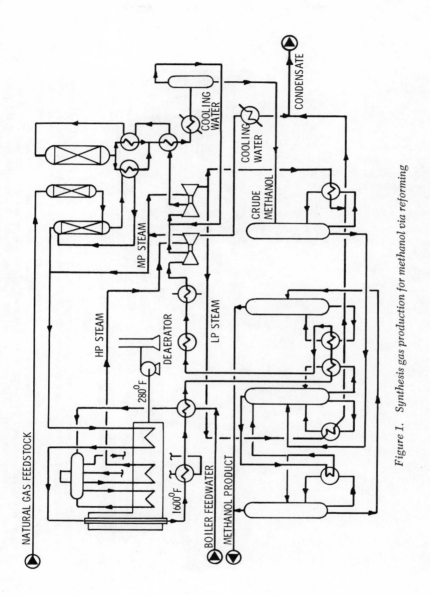

Figure 1. Synthesis gas production for methanol via reforming

Figure 2. Ugine Kuhlmann (France). Davy Powergas in 1971 designed through its French subsidiary, and in association with Technip, a new turnkey 660 TPD methanol plant at Villers St. Paul for Ugine Kuhlmann. The plant feedstock was Gruningen natural gas (containing 14% nitrogen). Synthesis occurs under low carbon concept conditions. Product methanol meets the US Federal Grade AA specification. Strict control of effluent quality is a feature of the plant. When this plant commenced production in 1973, Ugine Kuhlmann's methanol output was more than doubled and was greater than the total French production of methanol in 1970. The contract was for complete design and construction of the plant together with offsites. This was the first licensed plant to use 100 atm synthesis pressure.

Figure 3.   Georgia Pacific Corporation, USA.   This plant (on-stream August, 1971) produces 1000 STD of refined methanol to the US Federal Grade AA specification. The unit is based on natural gas feedstock and was the first plant operating under low carbon concept conditions, i.e., carbon dioxide addition to improve the stoichiometry of the synthesis gas is not required. This unit is based on centrifugal compression throughout and includes three distillation columns.

If additional CO_2 is available, it may be added prior to synthesis gas compression or it may be added to the reformer feedstock. Although the advantages of adding CO_2 are significant, the differential economics of the CO_2 addition points is marginal.

The synthesis gas is compressed to 80 atm in a two barrel centrifugal compressor driven by an extraction condensing steam turbine. The compressor usually has three stages of intercooling.

Methanol Synthesis. The compressed reformed gas is mixed with recycle gas from the methanol separator, and the combined stream is then partially preheated after passing through the loop circulator. The loop circulator is a high efficiency, single wheel, centrifugal compressor, driven by a backpressure turbine. The converter feed gas stream is then split into two streams. One stream is further preheated (in the warm shell interchanger) to the reactor inlet temperature. It passes to the converter via a Davy patented distributor. This distributor ensures even gas distribution over the catalyst bed even for extremely large converter sizes. As the gas passes down through the copper based catalyst, it reacts to form methanol, and generates heat according to the reactions below:

$$CO_2 + H_2 \longrightarrow CO + H_2O$$

$$CO + 2H_2 \longrightarrow CH_3OH$$

The catalyst bed temperature rise is controlled within a fairly narrow range by injecting portions of the second cool gas stream into the catalyst bed via specially designed lozenges. These gas distributors, which are submerged in the single continuous catalyst bed, are a special feature of the ICI process. They ensure complete mixing of the two gas streams while allowing extremely rapid catalyst changeout when required.

The hot converter effluent gas is split into two streams. Sufficient gas is passed to the warm shell interchangers to preheat the inlet gas to reaction temperature. The second stream is used to recover high grade heat to circulating water from the feedstock saturator. The two effluent gas streams then recombine and pass through the cold shell interchanger where remaining useful low grade heat is recovered to the incoming reaction gases. Crude methanol condensation begins to occur in this exchanger. Finally the effluent gases are cooled against cooling water where the majority of the crude methanol condensation occurs.

The two phase mixture then passes to the methanol catchpot where the two phases are separated. After an inerts purge is taken, the gas passes back to the suction of the loop circulator.

The crude methanol is let down to a lower pressure where the majority of dissolved gases are evolved. These and the loop inerts purge gases pass to reformer fuel system, after scrubbing with water to recover flashed methanol.

Methanol Distillation. To maximize heat recovery from the reformed gas, a four column distillation system is used to produce high purity product (Federal Grade AA).

The crude methanol from the crude methanol tank passes to the crude feed/topping column overhead exchanger, and then flows to the topping column where the light ends are removed. The heating medium of the topping column reboiler is LP steam. The light-ends purge from this column is used as fuel for the burners in the furnace.

The bottoms from the topping column are pumped directly to the refining column feed vaporizer and partially vaporized with reformed gas before entering the refining column. The reboil heat to the refining column is supplied by the refining column reboiler which is heated by LP steam. The purpose of the refining column is to remove the bulk of the water from the crude methanol and allow a purge for most of the heavy alcohols present. The refining column vapor overhead provides heat for reboiling the finishing column which operates at lower pressure. The overheads from the refining column are fed to the finishing column, where further removal of water and complete removal of the heavy alcohol impurities occurs. The majority of the refined methanol product is obtained from the finishing column. The bottoms of the finishing column, which contain predominantly water, methanol and higher alcohols, are fed to a small column called the polishing column where further methanol product is recovered. The water and higher alcohols, as well as a small quantity of methanol, are separated and taken from the bottom of the polishing column. All liquid purges taken from the distillation section are sent to the heavy alcohol tank from which they are pumped back to the reformer as feedstock or fuel. The water stream from the refining column bottom is pumped to the saturator and to the let-down vessel.

Advantages and Recent Advances in Methanol from Steam Reforming. Since the ICI LP methanol plant was first introduced in 1967, the design has been modified to enhance energy recovery as it became more economic to do so, due to increasing energy costs. Figure 4 shows this trend.

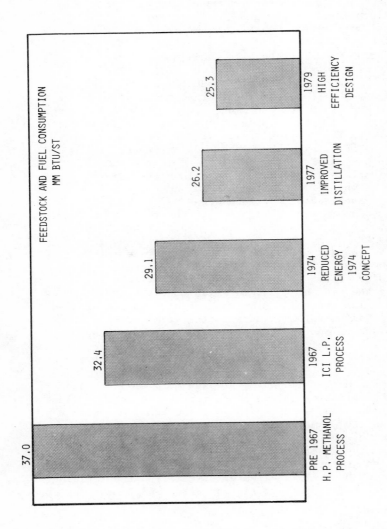

Figure 4. The continuing development of ICI's LP methanol process

Most of the efficiency loss in methanol production occurs in the reformer (]). This is because high grade fuel must be burned to supply the reforming heat load and combustion is a thermodynamically inefficient process.

The high efficiency design, therefore, depends upon recovering the excess heat, at as high a temperature as possible. This is achieved by total plant energy integration. The salient features of the high efficiency design are as follows:

Feedstock Saturator. The feedstock saturator provides over 49% of the process steam (60% with CO_2 addition). The saturator makeup includes the water effluent from the distillation system, which eliminates any liquid effluents from the plant, except blowdown from boilers, saturator or cooling tower, which are small flows.

The recycle of water from the distillation system also has the effect of reducing the demineralized water makeup requirement.

High Efficiency Reformer Design. Maximum feedstock and combustion air preheat is used to balance steam generation with steam requirements.

A 1500 psig steam system recovers reformer heat at the highest grade economical, depending upon the cost of fuel.

A high reformer exit temperature of 1616°F is made possible by high alloy tube materials such as Manaurite 36X and Paralloy. This leads to a reduction in methane slip, and an increase in the CO/CO_2 ratio. Both effects enhance the plant's efficiency and result in a reduction of feedstock natural gas consumption of 3.5% over previously used reforming conditions.

Highly Selective Synthesis Catalyst. Very low byproduct formation, and very high loop carbon efficiency is achieved by the use of ICI's 51-2 methanol catalyst, which is highly selective.

Maximum Heat Recovery. Maximum heat recovery in Davy designs is achieved by:
- judicious design of heat recovery trains to minimize degradation of heat quality,
- employment of a high thermal efficiency methanol loop with the recovery of over 2 mm Btu/Ton of high grade heat to process steam, and
- use of a four column distillation system to maximize the use of the low grade heat from the reformed gas that was previously rejected to cooling water.

Davy developed this system using the most advanced distillation computer technology available today. This

system has a twofold advantage over the two column system
previously used. Firstly, the total heat requirement of
the distillation section is reduced by 15%. Secondly and
more significantly, the distillation reboil duty uses
lower grade waste heat down to about 220°F, whereas the
old system could recover such waste heat only to about
300°F. Thus the effective savings amount to over 1 mm
Btu/Ton of reboil heat.

   Purge Gas Expander. A purge gas expander can be used
to drive either a major turbine or an electric generator.
This application is particularly suitable for large plants
and for non-stoichiometric loops.

   The economics of purge gas expansion is governed by
the cost of the expander which tends to be insensitive to
the HP delivered. These are sophisticated machines which
operate at high speeds in the order of 20-30000 RPM. They
are not economic below about 2 MW of power generation (2).
Thus the application is particularly suitable for large
plants operating on the Low Carbon Concept.

   Davy has installed two of these expanders, both for
the Celanese Chemical Company in USA and both are operating
successfully.

   Future Trends. A critical assessment of Davy's high
efficiency design shows that most heat is now rejected at
very low levels. Thus, further process development will
increase plant efficiency only slightly and will be
increasingly costly. Some marginal areas do exist, however, which may become economic as energy costs increase.

   Absorption refrigeration utilizes low grade heat and
may be employed to chill compressor suction and hence
reduce compressor horsepower. Similarly it may be used to
chill the loop circulating gas and increase the conversion
per pass. This would lead to a reduction in the circulator
power of up to 10%.

   Power recovery turbines, which let down suitable
refrigerants under pressure, are also being developed (2).
In conjunction with absorption refrigeration, these could
be used to recover power from compressor coolers and other
low grade heat.

   The Limitation of Present Technology. However, the
ability to increase efficiency by reducing plant drive
power requirements is limited by the reformer which is the
sole consumer of natural gas.

   Reducing plant drive requirements manifests itself in
a reduced demand for HP steam. This means less heat
recovery is required from the reformer flue gas duct. But
this will not lead to a reduction in reformer firing
because reformer firing is now governed by the radiant box

efficiency (there is no auxilliary firing). Thus a reduced requirement for HP steam would not lead to a reduced energy consumption on the Davy high efficiency design, unless alternate uses exist for the loop purge gases, or the export of power or steam to other process facilities is possible.

Natural gas reforming catalyst technology is well accepted throughout the industry. Davy offers the highest efficiency reformer design available today (over 94%), so we believe that using today's technology, it is not economically possible to reduce methanol specific energy consumption significantly below 25 mm Btu/St of refined product. This figure includes realistic heat losses, machine efficiencies, reformer air ingress, etc., which are achieved continuously in modern day plants.

Future Emphasis on Reliability and Capital Cost. We believe emphasis in methanol plant design will turn to reliability and capital cost.

In terms of reliability, the ICI LP methanol process is the undisputed leader, demonstrated by the following:
- there has never been a single commercial catalyst failure,
- a high onstream factor results from the consistant reliability of the ICI catalyst. Expected life of this catalyst is in the order of three years with some users reporting over five years' catalyst life. No other methanol catalyst has achieved this record.

We believe the next phase of development of the LP methanol process will relate to reduction in capital costs. This will become particularly significant as plant sizes increase. ICI is dedicated to continued further development of its catalyst to improve activity and life. This will lead to a reduction in the reactor size and catalyst volume and, consequently, to a reduction in capital cost.

Possible Long Term Developments. In the long term, significant further reduction in the energy consumption of the process can only be achieved by decreasing the radiant box heat load and reducing the steam for make up gas compression simultaneously. This will not be easy to achieve. Certainly, a new breed of high activity reforming catalysts and methanol synthesis catalysts will be required.

Reducing the tube sensible heat load is the only means of reducing the radiant section heat load. At present, about 28% of the tube heat load is sensible heat as opposed to reaction heat. The sensible heat load is minimized as the tube inlet temperature approaches the tube outlet temperature. This is most practically

achieved by reducing the tube outlet temperature which,
however, adversely affects the reforming equilibrium so it
is necessary also to reduce the reforming pressure, leading
in turn to an increase in synthesis gas compression requirements.

Alternatively, one might consider providing part of
the reformer heat load from a lower grade heat source,
such as the synthesis loop, rather than reformer fuel.
Unfortunately, reforming is favored by high temperatures
which suggests that supplying high grade heat by combustion
is the only way to supply the required reaction heat. On
the other hand, methanol synthesis is enhanced by low
temperature, which reduces the usefulness of heat liberated
from the reaction.

So, therefore, in all respects the two reactions are
incompitable. Even if new and more highly active catalysts
can be developed, it will be difficult to match them such
that a significant reduction can be achieved in the efficiency of methanol production via reforming.

Partial Oxidation of Hydrocarbons

Non-catalytic partial oxidation (POX) of hydrocarbons
from residual fuel oils to methane is commercially proven
by two processes, one offered by Texaco and the other by
Shell. Davy has experience with both processes. Each
process has a large number of plants in operation, with
feeds varying from natural gas to high sulfur residual
oil. (In fact, so long as the feedstock can be pumped, it
is a suitable feestock for a partial oxidation gasifier.
To this end, both Texaco and Shell are developing their
processes to handle coal slurries). The Texaco process is
somewhat more flexible in that it has commercial operating
experience with pressures up to 1200 psig. Thus the
Texaco process requires no synthesis gas compression. The
Shell process operates generally around 600 psig. Typically,
today's POX plant would be around 60% thermally efficient
(plus oxygen import) and can operate on cheaper feedstocks.

However, the economics of partial oxidation are
significantly affected by a requirement of about 0.9 tons
of oxygen per ton of methanol product.

The main advantage of partial oxidation lies in its
feedstock flexibility, so that the cheapest feedstock can
always be used.

Partial oxidation, although a major source of synthesis
gas in the 1940's, has gone out of favor in recent years
due to a shortage of heavy feedstocks and the lower cost
of steam reforming. However, due to its feedstock flexibility it will continue to play a growing although secondary
role in synthesis gas production in the future.

Partial oxidation is at present being adapted to coal gasification using a coal slurry as feedstock. There, in competition with other coal gasification processes, it may play a significant role, primarily because it is a commercially proven means of gasifying at high pressure. However, with coal gasification partial oxidation, economics are hampered by an increased oxygen demand caused by the slurry water addition. Whether the economics will be favorable is yet to be seen. Figure 5 is a typical process flow diagram of methanol production via partial oxidation of heavy fuel oil.

Brief Plant Description. Oxygen is reacted with a feedstock/steam mixture in a Texaco or Shell generator to produce a synthesis gas with a low H_2/CO ratio together with carbon dioxide, methane, sulfur compounds and a small percentage of unreacted carbon.

The synthesis gas from the generator passes through a waste heat boiler, raising high pressure saturated steam, and then through a boiler feedwater economiser before passing to a carbon scrubber, where carbon formed during the reaction in the combustion chamber of the generator is removed. Final traces of carbon are removed by scrubbing in a venturi device. In the Texaco process the carbon/water slurry is passed to the carbon removal plant where carbon is recovered and recycled to the generator. In Shell plants it is pelletized and burnt in a boiler.

A portion of the gas leaving the carbon scrubber is passed to a saturator/cooler unit where an increase in the steam/gas ratio is effected. From the saturator/cooler the gas passes to a single stage CO conversion section where CO is converted, in the presence of steam, to CO_2 and H_2.

The gas from the CO conversion section is mixed with the bypass un-shifted gas and is then cooled in a series of heat exchangers which supply heat to the distillation section and also to water preheaters for the steam system.

The synthesis gas from the waste heat recovery is rich in acid gases and is passed to a "Rectisol" or "Selexol" plant for their removal. During this process small amounts of hydrogen, carbon monoxide, methane, nitrogen and argon are dissolved in the solution and lost from the system in the acid gas stream.

The "Rectisol" or "Selexol" plant of the "selective" type absorbs both H_2S and CO_2, but in the regeneration section of the plant the H_2S and CO_2 are separated to provide two streams, one of which is CO_2, virtually free from H_2S, and the other is of a suitable CO_2/H_2S ratio to be processed in a conventional sulfur recovery plant.

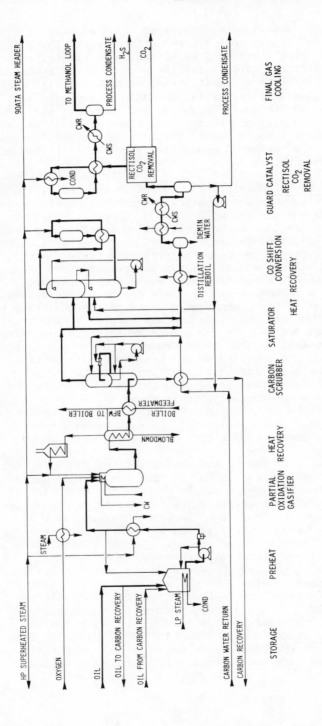

Figure 5. Synthesis gas production for methanol via partial oxidation

In order to prevent any sulfur or other poisons, such
as hydrogen cyanide, from reaching the methanol synthesis
catalyst, the make-up synthesis gas is passed over an
absorbent catalyst and finally cooled for the suction of
the synthesis gas circulator in the methanol synthesis
loop.

Coal Gasification

Coal, the foundation of the Industrial Revolution
until the Twentieth Century, will re-emerge as the major
energy deposit and chemical feedstock during the Twenty-
first Century.

There are three well proven processes for the gasi-
fication of coal. All three processes were developed in
Germany at a time when natural gas was not available in
Europe. One such process is the Winkler Process.

None of these current processes has a distinct
advantage over the other two in all situations of coal
gasification. However, we believe that the Winkler is
best suited for the production of methanol because it
yields a clean gas with a low inerts content. Other
advantages include the plant's low environmental impact,
the minimal coal preparation required and the simple
construction of the Winkler gasifier.

Brief Process Description. The coal is stored as
received in "active" or "inactive" coal storage area and
transferred to primary and secondary crushing facilities
where coal is crushed to 1/4" x 0 size. A portion of the
coal is sent to the coal fired boiler area and the rest is
fed through fluidized bed coal driers. The dried coal is
withdrawn continuously and transferred to the dried coal
storage silos. The dried coal is transferred to the
gasification section where it is gasified in a pressurized
Winkler gasifier. The raw gas generated from the gasification unit is cooled in the waste heat recovery unit and
is scrubbed with water in the particulate removal section.
The clean raw gas is partially shifted in the high temperature shift unit to produce the required ratio of carbon
oxides and hydrogen for synthesis. This gas is then
cooled and compressed. The compressed gas is sent to the
acid gas removal unit to remove sulfur compounds and the
required amount of CO_2. The purified gas flows to the
methanol synthesis loop.

The purge gas from the methanol synthesis unit contains
mainly methane. Depending upon the size of the plant,
this stream may be reformed to supplement the synthesis
gas produced from coal gasification.

The H_2S containing stream from the acid gas removal
unit is sent to the sulfur recovery unit to produce molten
sulfur. Figure 6 is a typical process flow diagram of synthesis gas production for methanol via coal gasificiation by
the Winkler gasifier.

Gasification of Wood

In certain countries of the world, trees grow at a
remarkably fast rate due to favorable climatic conditions.
If this country lacks significant energy deposits of any
other form, wood assumes an economically viable and totally
regenerable fuel source.

Such a country is Brazil. Burdened under a massive
balance of payments deficit due mainly to a lack of any
petroleum reserves, Brazilians have been looking to other
fuel sources for some time. Already some Brazilian gasoline
contains a substantial percentage of ethanol derived from
biomass fermentation (principally sugar cane bagasse and
molasses).

It was in this economic climate that a major Brazilian
company recently awarded to Davy Powergas the design of a
2000 TPD methanol plant based upon the atmospheric gasification of Eucalyptus Trees.

The methanol will be used to drive locomotives and to
fire thermal boilers in place of fuel oil.

Davy has, in fact, built over 20 biomass gasifiers
based upon feedstocks varying from cotton seed husks to
bagasse, but generally based upon wood.

Brief Process Description. Suitably sized wood
blocks are fed to a fixed bed gasifier operating essentially at atmospheric pressure. The gasifier is fired with
pure oxygen at a rate of about 0.4 tons/ton of methanol
product. After heat recovery, the gas is blown through a
Lymn washer where the tar and water soluble organics and
salts are removed by countercurrent contact with water.
The gas then passes through an electrostatic precipitator
before being compressed.

The compressed gas is shifted in a high temperature
shift converter and passed through a Rectisol CO_2 removal
system. In this way, a clean stoichiometric gas is obtained,
suitable for compression and injection into a standard ICI
LP methanol loop. Byproduct tars are fed to a partial
oxidation gasifier, adding to the synthesis gas supply.

The plant consumes about 3 tons wood/ton methanol
produced, based upon 35% moisture content of the wood.
The wood for the above mentioned Brazilian plant can be
grown on a regenerable basis from about 90 square miles of
forest (57,600 acres). Figure 7 is a block flow diagram
showing the many process steps involved in Davy's wood
gasification process.

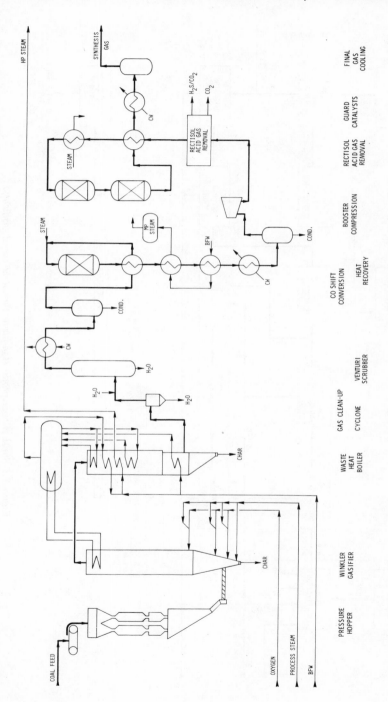

Figure 6.   Synthesis gas production for methanol via coal gasification

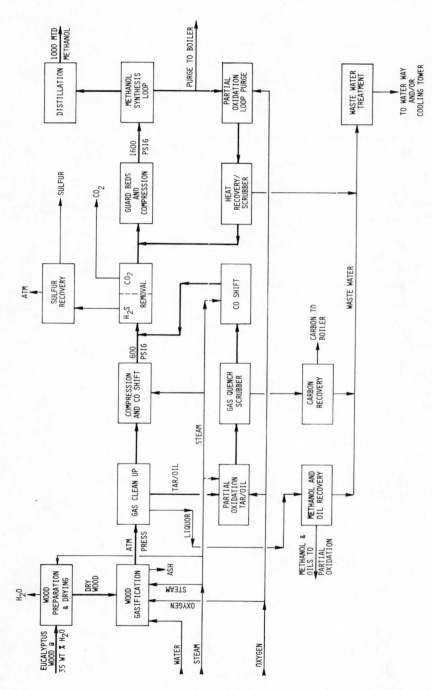

Figure 7.   Synthesis gas production for methanol via wood gasification

Feedstock Summary

Thus we have several different processes for the production of hydrogen and CO for methanol synthesis. The most important today continues to be steam reforming of natural gas and naphtha.

Coal gasification has been receiving much increased attention ever since the oil crisis of 1973. In the long term, it will become a significant source of methanol, because it will be the only nonregenerable energy deposit remaining. However, today it is competing with natural gas which is still available in large quantities and relatively cheaply.

In fact, today there is enough natural gas being flared to supply the world's total requirement of methanol. Coal's immediate future is in boilers which will free petroleum products for the transport and chemical markets.

Figure 8 shows the relative energy consumptions of these processes as currently offered by Davy Powergas using the ICI LP methanol process. These energy consumptions include realistic heat and other losses achieved continuously in modern day plants.

Current and Future Markets for Methanol

The current markets for methanol in the US are summarized in Table I.

Table I - Methanol Current Markets

End Use of Methanol	Millions of Tons 1976
Formaldehyde	1.205
Solvents	0.288
Chloromethanes	0.229
Methylamines	0.170
Acetic Acid	0.145
Methyl Methacrylate	0.140
Dimethyl Terephthalate	0.113
Glycol Methyl Ethers	0.041
Miscellaneous Chemical Uses	0.18
Miscellaneous Fuel Uses	0.075
Exports (Net)	0.142
Unaccounted	0.334
Total	3.12

Future Markets. Ammong the traditional markets for methanol, not much change is expected in usage and growth which will remain at about 6% to 7%. This will require the equivalent of a new 1000 TPD plant in US about every

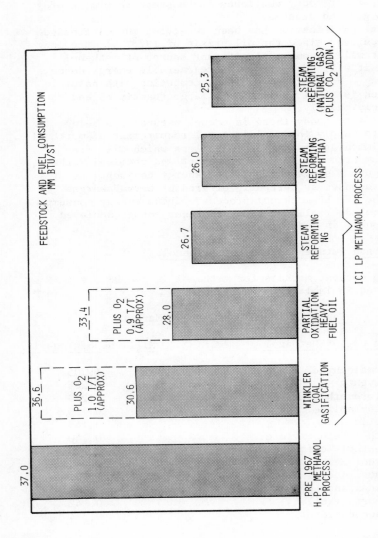

Figure 8. Current energy consumptions for methanol manufacture by various syngas production processes and feedstocks

two years. Similar growth is expected worldwide for
traditional chemical usage. One exception to this may
possibly be found by reviewing the history of methanol
usage.

During the 1950's methanol was replaced by ethylene
glycol as an antifreeze due to cost considerations. A
possible cost reversal during the late 1980's is conceivable, which would result in startling growth in methanol
demand.

However, it is in the possible future markets for
methanol that this commodity's major growth potential
lies. Figure 9 depicts methanols unique versatility in
todays energy picture.

Will Methanol Replace Gasoline? Perhaps the most
promising future use of methanol is as an automotive fuel
supplement. There are four ways in which this can be
achieved.

Firstly, pure methanol can completely substitute for
gasoline. However, this requires significant modifications
to existing vehicles including increasing the compression
ratio to about 14/1 for operation on lean mixtures and
replacement of some engine materials. (Certain elastomers
used for seals and gaskets, as well as aluminum and magnesium, are subject to attack by methanol).

Also, marketing of pure methanol would require a completely new distribution system, as was required in the US
when lead free gasoline was introduced.

On the positive side, 100% methanol produces about
10% more power than gasoline (hence the use of 100% methanol
at the Indianapolis 500 race for years). Also methanol
extends the lean misfire limit and gives slightly better
thermal efficiency than gasoline and, significantly,
methanol emits less nitrogen oxides than gasoline.
Studies carried out by General Motors have concluded that
100% methanol shows potential for meeting the ultimate NOX
emmission standard of 0.4 g/mile.

Thus 100% methanol does provide a viable alternative
fuel for automobiles, but, due to the significant modifications required to both automobiles and distribution
systems, is not likely to be introduced in the near future.

An exception to this may be the use in certain diesel
motor fleets, where methanol leads to a 10% increase in
power and gives cleaner emissions.

Methanol Blends with Gasoline. A more probable
development is the introduction of a methanol/gasoline
blend. This fuel requires no modifications to existing
car engines, and can use the existing distribution network.
It could therefore act as a stepping stone between gasoline
and 100% methanol.

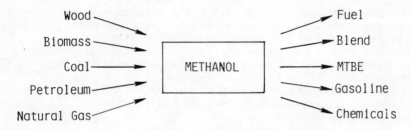

Figure 9.   Methanol—the versatile fuel

A blend of 15% methanol with gasoline (M15 blend) is the most practical initial approach because it requires only minor adjustments to engine fuel systems and is interchangeable with gasoline. It also provides a 3% improvement in energy efficiency, which largely offsets the lower heating value of methanol.

Several countries are already turning to methanol and higher alcohol blends to supplement gasoline. Indeed, Brazil, a country with no significant oil deposits, already uses a blend of ethanol in some of its gasoline.

New Zealand also has no significant oil reserves but large natural gas and coal deposits. The New Zealand government has committed the country to the use of methanol as a fuel supplement. This methanol will be manufactured from natural gas reserves and perhaps, in the long term, from coal.

In Australia, also, there exists an impending shortage of domestic crude oil, but vast reserves of natural gas are available in the Australian N.W. Shelf natural gas deposits. It has been shown in some studies (3) that M15 synthesized from domestic natural gas is the most economic solution to the crude oil shortage in Australia.

Octane Boosters. A third method of supplementing gasoline with methanol is via the gasoline additive methyl tertiary butyl ether (MTBE).

MTBE is synthesized by a simple process in the liquid phase, over a solid ion exchange resin which catalyses the reaction of methanol with isobutylene.

MTBE is primarily an octane booster but, unlike existing octane boosters, it contains no metals. MTBE is already used in Western Europe as a fuel additive and was recently approved by EPA for use in gasoline in USA.

Total domestic capcity of MTBE in the US is, however, limited by the availability of isobutylene. The most available source of isobutylene is from ethylene crackers and if all from that source were utilized to produce MTBE this would yield only 50,000 bbl/day of MTBE. All available isobutylene would produce 200,000 bbl/day of MTBE. At blends of 7%, this would only satisfy the existing unleaded gasoline market, but not that being opened up by the phasing down of lead in gasoline scheduled for October of this year.

Gasoline from Methanol. Mobil has developed a process, now in the pilot plant stage, which dehydrates methanol to gasoline producing a 93 research octane gasoline.

This process offers a possible source of gasoline from coal at a cost not significantly above possible future Middle East gasoline prices. It is seen princi-

pally as an alternative to Fisher-Tropsch synthesis of
liquid fuels, and the relative economics of both
approaches are under active study.

Methanol a Versitile Fuel. Thus there is no doubt
that methanol will assume a significant role as an automotive fuel. At present its application is limited to
situations where local conditions fit the advantages of
methanol as a fuel, in any of its guises.

In the future such situations will increase as petroleum
reserves become less abundant and manipulated more politically.

Other New Markets. Moving away from the automobile
fuel market, ICI and Phillips Petroleum have developed
single cell protein (SCP) products made by the growth of a
microorganism that uses methanol as the sole source of
carbon nutrient. These SCP's aim at the animal feed
market as high protein additives in competition with
skimmed milk powder and fish meal. If this process proves
successful it offers a potential future additional consumption of methanol, although not in the same league as
the possible fuel market.

In conclusion, it can be stated that there is a high
probability that methanol production will increase world-
wide by an order of magnitude, as the advantages of its
ease of production from cheap fuel sources, its beneficial
effect upon the environment and its versatility are economically realized in the next twenty years.

Literature Cited

1. A. Pinto and P.L. Rogerson, "IMPACT OF HIGH FUEL COST
    ON PLANT DESIGN" Chemical Engineering Progress, July,
    1977.
2. G.C. Humphreys, J.R. Masson and M.J. Pettman,
    "DEVELOPMENT OF METHANOL PLANT DESIGN FOR LOW ENERGY
    USAGE" Ampo 78.
3. J.R. Bradley "METHANOL AS A GASOLINE SUPPLEMENT"
    Conference on Alcohol Fuels, Sydney, August, 1978.
4. E.E. Graham, B.T. Judd "EXPERIENCE WITH METHANOL-PETROL
    BLENDS" Conference on Alcohol Fuels, Sydney, August,
    1978.

RECEIVED October 10, 1979.

Technical and Economic Advances in Steam Reforming of Hydrocarbons

R. G. MINET

KTI Corporation, 221 East Walnut Street, Pasadena, CA 91101

O. OLESEN

United Technologies Corporation, South Windsor, CT 06074

For several decades, the process of high temperature steam
reforming (HTSR) has been the most efficient, economical and
practical technique available for conversion of light hydrocarbons
to hydrogen and hydrogen/carbon monoxide mixtures for eventual
manufacture of ammonia, methanol, oxo-alcohols, pure hydrogen
and a wide range of petrochemicals, chemicals and chemical
treatment atmospheres. Raw materials used range from natural
gas, methane, and methane containing refinery gases through
various combinations of light hydrocarbons including ethane,
propane, butane, pentane, light naphtha and heavy naphtha (420F
end point). Systems have been developed to remove various impurities, principally sulfur compounds, from the feedstock to
limit poisoning of the HTSR catalyst which is usually nickel based,
promoted with various combinations of alkalai and exotic metals.
The reforming reaction typically is carried out with steam to
carbon ratios in the range of 2.5 to 5.0 at process temperatures
from 1300 to 1600F and pressures up to 500 psig. Since 1970,
significant advances have been introduced in the design, fabrication and operation of HTSR equipment and systems which extend
the operating range, improve efficiency of use of feed and fuel,
reduce capital cost, and in the foreseeable future will extend
the technique to include distillate fuel oils as an acceptable
feedstock on a commercial basis. Many of the innovations are
essentially extensions of the state of the art which permit
improved efficiency by more complete heat recovery, closer
temperature approaches, and use of superior materials of construction. Some recent advances involve new process steps, new
catalyst and new process engineering ideas.

In addition to presenting a brief review of recent improvements, innovations and design approaches, this paper will describe
an advance design reformer developed originally by United
Technologies Corporation using space age technology concepts
which will soon be placed on-stream to produce hydrogen for fuel
cell use in the generation of electric power (1). Adoption of
the steam reformer in this system in commercial HTSR facilities

0-8412-0522-1/80/47-116-147$07.25
© 1980 American Chemical Society

will significantly alter the size and improve the efficiency, and
flexibility of this class of equipment in future process plants
making hydrogen from hydrocarbon raw materials.

Conventional Design

A wide variety of basic mechanical designs and arrangements
have been used for conventional high temperature steam reforming
reactors. All furnaces include fuel combustion systems, a radiant
heat transfer section where the high temperature required is
attained within the reaction tubes, a convection section where the
hot gases are cooled to recover the available heat and various
auxiliary systems required to supply combustion air, control the
combustion process and integrate the furnace with the process unit
it serves. There are advantages and disadvantages to all possible
flow schemes. By way of illustration, Figure 1 shows several
of the most common arrangements which are also summarized in
Table 1.
Various combinations of these configurations are in service
in facilities designed to produce hydrogen, hydrogen and carbon
monoxide mixtures for ammonia synthesis gas, methanol synthesis
gas and carbon monoxide/hydrogen mixtures for the production of
a wide range of chemical products ranging from simple alcohols
to complex organic chemical molecules.
Since the mid 1970's, changes in the design of reformers have
produced improvements in overall thermal efficiency, accomplished
primarily by improved heat recovery in the convection section,
ultimately reducing the temperature of the flue gas to the limit
imposed by the condensation point of water, or in the case of
some fuels, sulfur trioxide. Extensive use of air preheat in
newer designs lowers fired fuel requirements.
Fuels utilized in steam reformers have included various
gaseous streams ranging from natural gas to hydrogen-rich fuel
gases, process gases of various kinds, purge gases, coke oven
gas and various light liquid fuels from naphtha to the No. 2
distillate. The vaporized fuel oil technique licensed by Allied
Chemical Company has permitted an extension of the range of
liquid fuels which can be handled in radiant burners (2). Fuel
oils heavier than No. 2 have been utilized for firing reformers
but these are limited to low sulfur and metal contents by the
metallurgy of the reformer tubes and convection section.
Improvements in overall thermal efficiency have been coupled
with the introduction of extensive air preheat to decrease the
fired fuel requirements. Additional decreases in fired fuel
requirements can be obtained by materially improving the radiant
box efficiency through changes in radiant box geometry. Usually
a balance was struck between the fired fuel requirement to meet
the thermal reactor duty, and the overall need for additional
heat for generation of high pressure steam required for the
process as a reactant or for driving pumps, compressors, or

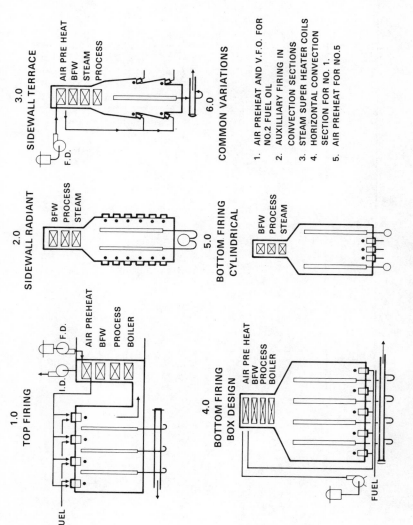

Figure 1.   Common arrangements for HTSR furnaces

Table 1. Common Arrangements for Commercial HTSR Furnaces

Item	Type				
	Top Fired	Sidewall Radiant	Sidewall Terrace	Bottom Fired box	Bottom Fired cylinder
Shape	rectangular box	long and narrow	long and narrow	rectangular box	vertical cylinder
Tube Arrangement	several parallel	one or two rows	one or two rows	several parallel rows	circular around periphery
Convection Section	at grade or vertical alongside box	elevated above radiant section	elevated above radiant section	elevated above radiant section	elevated above radiant section
Draft	forced and induced	natural or forced	natural or forced	natural or forced	natural or forced
Fuel	gas or liquid	gas or vaporized liquid	gas or liquid	gas or liquid	gas or liquid
Air Preheat	normally provided	possible	possible	normally provided	possible

blowers. A typical schematic presentation of the overall flow scheme in a modern high temperature steam reformer loop is given in Figure 2 for a 10 million scf/day hydrogen plant.

Using the materials of construction and mechanical design procedures which are typical of the late 1970's, a conventional high temperature steam reformer would have operating and design parameters approximately as given in Table 2 for a 10 million scf/day hydrogen plant.

Mechanical and Process Innovations (Sections A - H)

Designers of HTSR equipment continuously introduce new technology to reduce cost, to improve efficiency and to increase the range of feedstocks and fuels which can be handled. In today's economy, it is clear that very high thermal efficiency coupled with reasonable flexibility in feedstock and fuel characteristics is required for a viable hydrogen production system. Improvements in technology introduced in the last few years deserve detailed study and evaluation for any current design project. The following paragraphs provide an indication of the effect of the improvement on the overall HTSR design.

Burners - Section A

Various processing schemes involving steam reformers require use of fuels having a very wide range of combustion properties. At startup, a steam reformer may burn natural gas, LPG or No. 2 fuel oil. When the system is on-stream, at least part of the fuel may be a low heating value gas generated by a process recovery system, as for example, Pressure Swing Adsorption (PSA) purification, or it may be a purge gas from ammonia or methanol synthesis. Such a gas may have a heating value as low as 90 to 150 Btu/cubic foot. In some cases, the reformer may be fired at some time during its cycle, utilizing high hydrogen off gas from a refinery or from some outside process step producing an excess of fuel. A modern, well designed reformer must accommodate variations in fuel composition without loss of firing efficiency or disruption of the control system.

Multiple gun burners, piped up to handle the anticipated range of fuels for the particular operation are being incorporated in finely tuned steam reformers currently under construction. A diagram showing schematic details (3) for such a burner is shown in Figure 3. Newly available burners can handle both low and high heating value gaseous fuels with a liquid fuel as well while remaining within permissible limit for noise and NO_x restrictions. Such burners can be used for natural draft and forced draft installations, with or without air preheat.

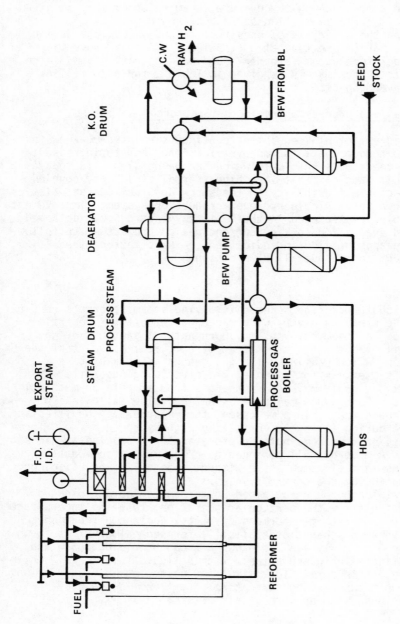

Figure 2. Primary and secondary shift convertors—HTSR loop for hydrogen plant

Table 2. Typical Design Parameters for HTSR for 10 MM scf/d Hydrogen Production

Feedstock	Natural Gas
Process flow, lbs/hr	29000 including steam
Inlet pressure, psig	400
Outlet pressure, psig	365
Operating temp. in F	1000
out F	1550
Maximum tube wall temperature	1625
Heat absorbed	40 Million Btu/hr.
Heat released	83 Million Btu/hr.
Tubes, No. Material	60, HK 40
Heated length	34 feet
i.d./o.d.	4. inches / 5.5
"as cast" thickness	0.75 inches
Heated surface area	2125 square feet
Catalyst volume	180 cubic feet
Flux based on inside diameter	18900
Burners	24
Air preheat temperature	500°F
Flue gas outlet, F	315

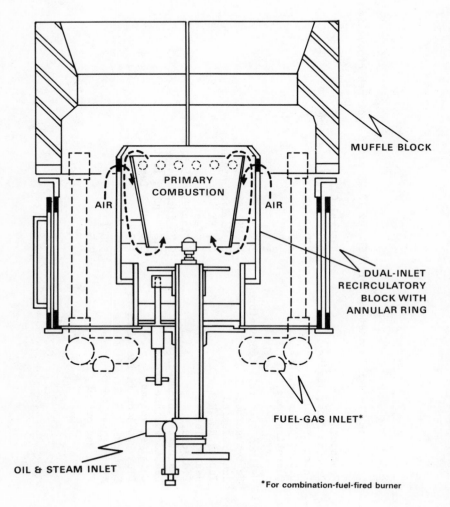

Figure 3. Multiple fuel burner reaction zone

Insulation - Section B

A fairly recent shift in the construction of reformer furnace equipment has been the widespread use of ceramic blanket insulation to replace the time-honored and well-proven refractory brick wall lining. In early applications, ceramic fiber material was utilized for repairs of brick and for the archway inside the furnace which was traditionally a source of difficulty. Use of ceramic fiber walls in top fired and cylindrical reformers has significantly improved their performance in terms of time required for startup and shutdown. This type of construction permits a much increased degree of prefabrication and modular shipping. Typical comparitive properties for brick and ceramic fiber lining are given in Table 3 to illustrate the qualities of materials involved. The lower required weight of ceramic fiber linings gives an added bonus in reduced steel cost as well.

Reformer Tubes - Section C

The most essential element of the steam reformer is the reaction tube in which the catalyst is contained. Typically such tubes are manufactured from very high alloy materials containing 25/20 to 25/35% chrome nickel plus various alloying metals including niobium, tungsten, and other selected elements to enhance the physical and chemical properties. The true strength and performance of the tube material depends on the integrity of manufacture. Specification for the high alloys are classified as HK 40 and HP 50. Typical composition for these alloys are given in Table 4. Furnace designers typically have rigid requirements for these materials in terms of chemical composition and test procedures. The reactor tubes from these metals are formed by a process known as "continuous centrifugal casting".

New techniques for the formation of high temperature tubes include a proprietary process called "weld forming" which has been developed and placed in practice by Mitsubishi Heavy Industries Ltd. This technique is reported to give improved micro structure in the metal and potentially higher allowable stress at the higher design operating temperatures (4). Industry practice normally limits the thickness of reformer tubes to less than one inch, thus for the average four inch tube diameter, the maximum allowable internal operating pressure for a process outlet temperature of 1550F is of the order of 500 psig. Use of new production techniques may permit some increase in both allowable thickness and allowable stress which could, after suitable tests have been completed, result in raising permissible operating pressures by some significant percentage. As a minimum, the improved metal structure should increase the actual service life of the reformer tubes from the average value of three to seven years towards the target goal of 80,000 to 100,000 hours.

Table 3. Comparison of Refractory Brick and Ceramic Fiber Properties

	Brick	Ceramic fiber
Basis outside wall temperature	150F	150F
Inside wall temperature	2100F	2100F
Thickness	3 inches block 2 inches firebrick	6 inches insulating fiber 6 lbs/cf 2 inches fire fiber 8 lbs/cf
Conductivities Btu/hr/S.F./in.°F	Block 0.7 6 lb/cf Firebrick 1.7 8 lb/cf	0.70 1.65
Density, lbs/cf	Block 18 Firebrick 31	6 8
Weight, lbs/S.F. of insulation, surface	28	5

Table 4. Typical Composition of Tube Materials for HTSR

Designation Type	HK 40		HP 50	More
manufacture	CCT	WFT	CCT	CCT
chemical composition weight %				
C	0.40	0.40	0.40	0.50
Si	1.00	1.00	1.20	1.75
Cr	25	24.5	25	25
Ni	20	21.5	35	35
other	—	Mn	nb	W
Service Temp °C				
Strength	1000	1000	1050	1150
Hot gas oxidation	1100	1100	1100	1200

CCT - centrifugal cast tube
WFT - weld formed tube

Convection Systems - Section D

In the design of the convection system of HTSR equipment, the important operations carried out are air preheat, boiler feedwater preheat, feedstock preheat and steam generation. The limits of thermal efficiency for these various operations are, to a certain extent, fixed by the mechanical design of heat exchangers, available materials of construction, allowable pressure drop and cleanliness of the flue gas. Looking at the low temperature end of the reformer package, the temperature approach which limits the overall efficiency is at the air preheater or boiler feedwater preheater and the flue gas entering the stack. The colder the stack gas, the more heat energy recovered and returned to the process. As the temperature of the stack gas decreases, it is obvious that the risk of condensation, with its related problems of corrosion due to acid gases, is the limiting factor. Glass covered tubes or glass itself has been used as heat exchanger material in the flue gas cooling system. Regenerative type air preheaters fabricated from enameled or stainless containing basket elements are available and have been used with some success for this purpose as well. Such systems frequently include on-line washing connections (5).

A relatively new development which promises significant economies in design cost and operate efficiency is the heat pipe. These compact elements contain a sealed in heat transfer medium which transfers heat from one end of the pipe to the other by either vaporization and condensation or by capillary action (6). Such devices promise less complicated ducting, lower pressure drop and closer temperature approach than the traditional systems using heat exchangers of the flow through type, or regenerative systems involving revolving baskets of heat transfer elements. A typical arrangement of one type of heat pipe available in today's markets is shown in Figure 4. Use of such elements may permit savings in investment and operating costs.

Catalysts - Section E

New catalysts which extend the range of feedstocks which can be successfully reformed in the catalytic reactor are being studied continuously. Systems generally utilize nickel in one form or another, promoted by various metal salts to enhance the endothermal reformer chemical reaction. Catalysts recently described in the literature as being available for commercial use or for further development are reported to be capable of operating successfully in high temperature steam reforming reactions of hydrocarbon feedstocks in the No. 2 fuel oil range or heavier as well as with coal-derived liquids, even those containing up to 0.4% sulfur. These catalysts seem to require higher operating temperatures for reasonable conversion of the

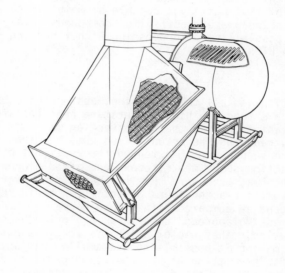

Figure 4.   One type of available heat pipe

feedstocks involved.

One such catalytic system has been pilot planted by Toyo
Engineering Company (7). It utilizes a combination catalyst and
specially designed feedstock vaporization system to process hydrocarbons as heavy as crude oil to obtain reasonable conversions
to hydrogen and carbon monoxide. Another process announced by
Grand Paroise uses a fluidized bed of catalyst for the reforming
step (8). Heat required is introduced into the fluidized catalyst
by burning fuel inside of tubes immersed in the bed. Both of
these systems have been extensively tested in large pilot installations and could be included in commerical installations
in the near future if justified by economic considerations.

Along with the new development in reforming catalyst, the
availability of shift catalyst capable of maintaining a long
active life in the presence of sulfur compounds have greatly
enhanced the possibility of processing higher boiling range
feedstocks in HTSR systems (9).

Process Design - Section F

In addition to all the foregoing, designs have been
proposed which extend the high temperature heat transfer function
from the radiant firebox into the convection sections, thus
significantly reducing the actual fired duty required for the
reformer. One such approach extended the reactor tubes into the
convection section above the firebox, and added finned surface
to improve the heat transfer. By this means, the overall heat
transferred to the reactor tubes directly can be increased to
52 to 54 percent as compared with a more normal 45 to 48 percent.
Another interesting approach described in a patent assigned
Fluor Corporation, makes use of adiabatic reactors external to
the furnace, making it possible to use smaller radiant tubes,
simultaneously permitting higher operating pressure and improved
thermal economy (10). This design is shown schematically in
Figure 5.

By using more of the high temperature heat available in the
flue gas and reaction products, the fired duty required can be
decreased by a significant fraction as compared with the more
conventional arrangements. Such concepts are not necessarily
new but they have become more interesting as the cost of fuel
and feedstock have escalated.

Pressurized Combustion Reformer - Section G

Recently, a completely new approach to the design of HTSR
furnaces was placed in pilot plant operation based on process
and mechanical design developments by United Technologies Corporation (UTC). This new reformer will be part of the fuels
processing system in an electric power installation designed
to produce 4.8 MW for the Consolidated Edison Company in

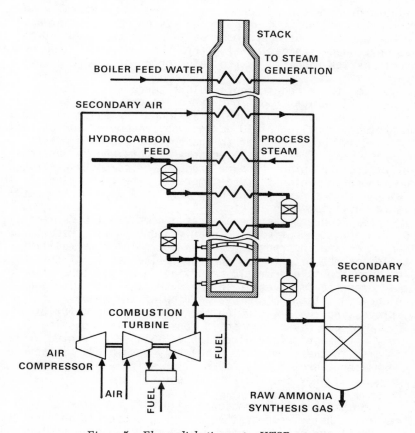

Figure 5.   Fluor adiabatic reactor HTSR concept

New York City using hydrogen derived from naphtha to feed a fuel cell power generation unit (11). The need to meet certain specific requirements forced a completely new approach to the steam reformer package. To attain the objectives of their design program, UTC was required to build a steam reformer which could be completely shop assembled, shipped in one piece by readily available truck or rail transportation. weigh no more than 20 tons, have no maximum dimension greater than 12 feet, and be capable of generating hydrogen and carbon monoxide from any available hydrocarbon raw material ranging from natural gas through 420F end point naphtha. The system has to start up burning feedstock but operate normally using the low heating value, hydrogen-rich gas exiting from the fuel cell. The overall characteristics required of the reformer assembly are given in Table 5.

To meet these requirements, UTC altered the basic reformer furnace design which traditionally relies on radiant heat transfer to the reactor tubes. By burning fuel under pressure inside a high pressure shell, the UTC design uses a combination of radiant and convective heat transfer to the reactor tubes from hot combustion gas. Overall heat economy is obtained by passing the high temperature, high pressure products of combustion through a turbo compressor which in turn compresses the combustion air for the system to the required 50 psig level. A view of the resulting pressure shell is shown in Figure 6.

The UTC concept was operated as a complete 1.0 MW facility at S. Windsor, Connecticut for nearly 3,000 hours. Results of this program are incorporated in the design of this facility installed at the site in New York City. The result of this design approach is a dramatic reduction in size, along with an equally dramatic improvement in the overall flexibility and efficiency for steam reforming. A view of the model of the fuel processing train including the reforming, desulfurizing train, turbo compressor and shift convertors is shown in Figure 7 with appropriate dimensions. This skid mounted system which comes to the plant site completely assembled, is capable of generating approximately 4.2 million scf/day of contained hydrogen as it provides the requirements of the 4.8 MW fuel cell system.

For comparison in dimensions, Figure 8 shows two of the UTC reformer shells representing a hydrogen production capacity of 6 to 8 million scf/day superimposed to scale on a photograph of a conventional top fired reformer with the same total hydrogen capability.

The advanced reformer unit combines naturally with a Pressure Swing Adsorption (PSA) system to permit supply of complete hydrogen facilities totally skid mounted and factory assembled with up to 6 million scf/day capacity in a single reformer train. A typical flow diagram for such a system is shown in Figure 9. Typical operating parameters when processing hydrocarbon

Table 5. Design Constraints for HTSR System 4.8 MW Power Generator

Dimensions   Maximum   12 feet
                          Weight   40,000 pounds

Fabrication   Complete in shop

Transportability – by truck or rail
                  must withstand force of 4.0 G's

Process

   Feedstocks   Natural gas, LPG, naphtha

   Efficiency   High

   Start up time   4 Hours from cold

   Flexibility   10% to 100% in minutes

   Safety   Operate unattended

   Maintenance   Every 16000 hours

   Environment   Low NO_x, low noise

*Figure 6.   UTC pressurized reformer hydrogen generation system model for 4.8
MW facility*

Figure 7.   A 6 × 10⁶ scfd hydrogen reformer and two 3 × 10⁶ scfd UTC hydrogen reformers

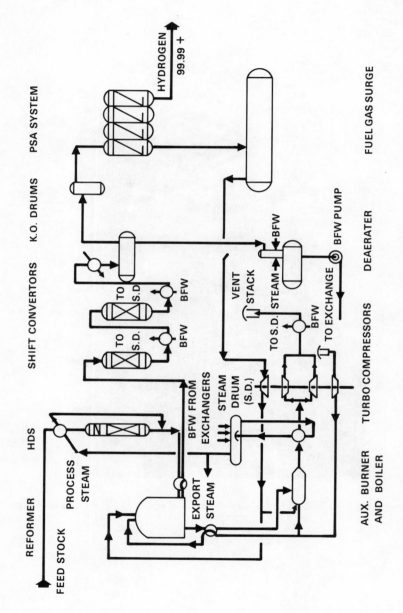

Figure 8. Hydrogen plant simplified flow sheet UTC reformer–PSA (KTI Corporation design)

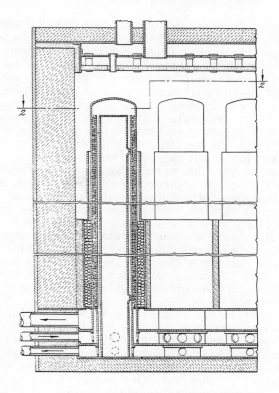

Figure 9A.   UTC reformer internals

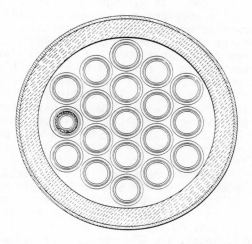

Figure 9B.   UTC reformer tube arrangement

feedstocks ranging from natural gas through naphtha are given
in Table 6. Such efficiencies in the range of 380 to 400 Btu of
feed and fuel per thousand standard cubic feet of hydrogen
produced are significantly better than the performance normally
attainable with conventional reformer design.

The indicative internal arrangement of the reformer as
revealed in the patent literature (12) is shown in Figure 10.
This design permits far greater utilization of the combustion
volume and overall reformer envelope than is possible using
conventional radiant box systems, leading directly to real
economies in investment and operating cost.

Examination of the flow pattern in Figure 10 shows that
combustion takes place in a carefully designed plenum and grid
assuring distribution of the hot combustion gases over the entire
reaction tube cross section. The tubes are arranged in a closely
spaced configuration with process gas flow moving entirely upward
through an annular space containing catalyst, then down through an
internal annulus which permits preheating of the inlet feed stream
while cooling the reactor effluent. The internally arranged
heat exchange makes it unnecessary to supply a high temperature
process gas boiler at the reformer outlet, eliminating a frequent
source of maintenance.

The basic restriction requiring a total shop fabrication
would limit the maximum capacity of a single reformer shell
following this design to something in the range of 8-12 million
scf/day of hydrogen. Hydrogen plants based on UTC reformers
could serve the food processing industry, direct ore reduction
facilities, synthetic fiber and other chemical and petrochemical
applications. The very compact reformer design provides a simple
way to add to existing capacity where reformers are the bottleneck
in hydrogen plants. Complete hydrogen generating facilities can
be shipped as totally prepiped and premounted units. Such systems
may be part of power generation fuel cell installations throughout
the USA and elsewere before the end of the 1980's.

Since the new design is actually a true modular unit, a
number of reformers could be combined to provide high capacity
HTSR installations with potential for economies in both material
and operating costs. By way of illustration, a conceptual design
for a reformer to serve a nominal 1000 ton/day ammonia or
methanol plant consisting of eight 8.5 million scf/day units is
shown in Figure 11. Such a system assembled with prepiped mani-
folds and turbo compressor can be installed in a fraction of the
time required for conventional top fired or side fired designs.
The combined modules shown in Figure 11 would measure approximately 32 feet by 60 feet on the base and 14 feet high overall.
A conventional top fired reformer would require a base area of
60 feet by 65 feet for the radiant box while the top fired burner
level would be at least 60 feet above ground level.

As previously noted, the UTC system operates with pressurized
combustion inside the shell of the reformer. One of the

Table 6. UTC-PSA Hydrogen Plant Typical Operating Requirements

Basis: One million std. cu. ft. per day hydrogen

Purity of H_2,%	99.99+
Water	less than 2 ppm
$CO + CO_2$	less than 10 ppm
Feedstock and Fuel	385 to 400,000 Btu per Mcf
B.F.W.	4 to 8 GPM
Export Steam	0 - 2000 pounds/Hr
Cooling Water	100 - 200 GPM
Electric Power	20 - 40 KWH
Delivery Conditions	200 psig
	100 F
Plot Area	50 Ft X 50 Ft

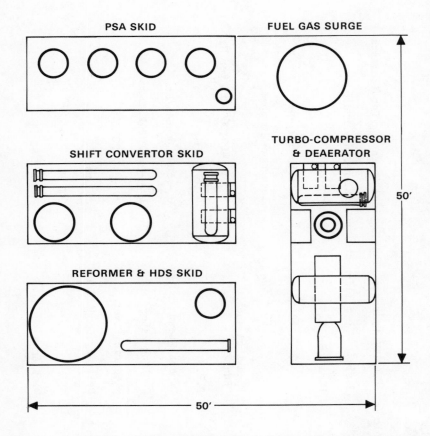

Figure 10.   UTC reformer–PSA hydrogen plant (schematic plot plan—KTI Corporation design)

limitations of HTS reforming systems in use to date is the
process side pressure which can be accommodated and controlled
by the metallurgy and permissible thickness of the reformer
tubes. The maximum process side pressure in use today is in the
range of 500 psig. By operating the combustion side at an
elevated pressure level, it is possible to project to reformer
systems which can operate with process side pressures at the level
required for methanol synthesis or eventually, certain types of
ammonia synthesis utilizing advanced catalytic materials. While
such developments are not available at this time for commercial
use, design studies made to date by KTI have uncovered no serious
obstacle to their realization.

Reforming Distillate Fuel Oils - Section H

The extension of the range of usable feedstocks to include
distillate fuel for high temperature steam reforming is under
serious and potentially successful development. The objective
is to provide a process which can convert commercially available
No. 2 fuel oil, having sulfur contents in the range of 0.4 weight
%, to hydrogen and carbon monoxide mixtures for production of
ammonia, methanol and hydrogen with the overall efficiency and
economy attainable with the high temperature steam reforming
process for light hydrocarbon feedstocks.

The approach being used to achieve this result was described
recently by KTI reporting on work carried out for the Electric
Power Research Institute (13). The basic scheme, in block
diagram form, is shown in Figure 12. It consists of a hybrid
reactor combining an advanced tubular steam reformer with an
autothermal catalytic reformer. This combination overcomes the
limitations of lower catalytic activity in HTSR systems toward
the heavier hydrocarbons while retaining a significant part of
the desired process characteristics.

Development of the hybrid reactor technique is entering
a bench scale experimental program, to be followed with a small
scale prototype unit scheduled for operation late in 1979.

Summary and Conclusions

Advances in the technology of high temperature steam reforming are being made continuously to provide improvements in
efficiency, feedstock range and operating parameters. A review
of recent advances gives unmistakable evidence that vigorous
development activity with fresh view points will materially improve cost and operating factors for this extremely useful process
technique.

Application of the new heat recovery techniques, mechanical
materials of construction and combustion systems to conventional
reformer furnace design will help to keep the HTSR even with
material and feedstock cost escalation. Adoption of the

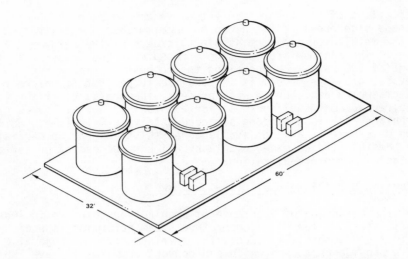

Figure 11. Steam reformer modules conceptual arrangement for 1000 tons/day ammonia plant

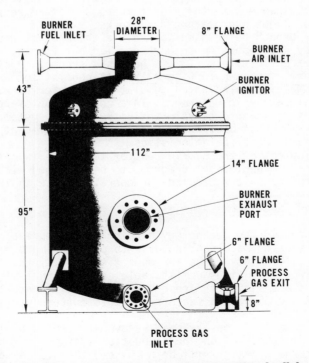

Figure 12. Steam reformer under construction for 4.8-MW fuel cell demonstrator power plant (to be operated by Con Edison in New York City)

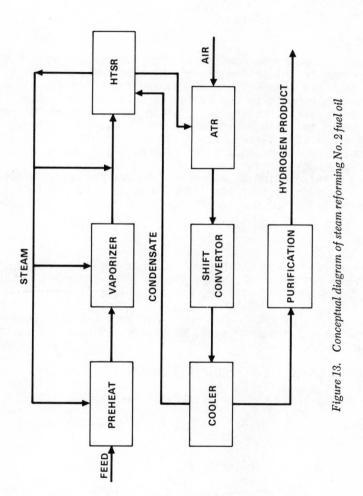

Figure 13.   Conceptual diagram of steam reforming No. 2 fuel oil

pressurized combustion reformer demonstrated for fuel cell
installation in 1978 and 1979 will lead to real improvements
in the efficiency and capital cost requirement of hydrogen generation facilities to give benefits of sufficient magnitude
to have an important impact on the course of the synthetic
ammonia, methanol and hydrogen process industries.

Abstract

     Steam reforming is now and for the foreseeable future the
most economic technique for converting hydrocarbons ranging from
methane to heavy naphtha to hydrogen and synthesis gas. Significant developments in the design of reformers and associates
systems have taken place recently which have extended the range
of feedstocks to include distillate fuel oil, and provide
interchangeability of feed, rapid response to change in demand
while providing for production of high purity hydrogen with
remarkable economy of feed and fuel.
     Proper integration of advanced reformer design with adsorp-
tion purification systems can reduce overall feed and fuel
requirement below 400 Btu per standard cubic foot of hydrogen.

References

1.   Sederquist, R. A., "Evolution of Steam Reformers for Commercial Fuel Cell Power Plants," Symposium on Reformers and Hydrogen Plants, March 7-9, 1978. Santa Barbara, California

2.   "Application of VFO Technology," Hatcher, D.C., Symposium
on Reformers and Hydrogen Plants, March 7-9, 1978, Santa Barbara,
California.

3.   National Air Oil Burner Company, Bulletin

4.   Mitsubishi Heavy Industries, Ltd, Bulletin

5.   Baker, F. J. and Bloomfield, K. F., Oil and Gas Journal,
December 4, 1978

6.   Q-Dot Corporation Bulletin

7.   Tomita, T. and Kitogawa, M. "New Steam Reforming Process for
Heavy Hydrocarbons", Achema 76, Frankfurt, West Germany, 1976

8.   Chemical Week, May 12, 1976, P. 71-75

9.   Chemical and Engineering News, June 28, 1976

10. Bogart, M. J. P., Hydrocarbon Processing, April 1978, P. 145-151

11. Handley, L. M., Rogers, L. J. and Gillis, E. "4.8 Megawatt Fuel Cell Module Demonstrator," August 1977 report based on contract No. Ex-76-C-01-2102, U.S., E.R.D.A., E.P.R.I. and U.T.C.

12. U.S. Patent No. 4,098,587, July 4, 1978, Krar, G. R., Olesen, O., Sederquist, R.A., and Szydlowski, D.F.

13. Kinetics Technology International Corporation, December 1978. "Assessment of Fuel Processing Systems for Dispersed Fuel Cell Power Plants", prepared for Electric Power Research Institute, California

RECEIVED September 21, 1979.

Coal Gasification for Hydrogen Manufacturing

W. G. SCHLINGER—Texaco Research Center, P.O. Box 400, Montebello, CA 90640

J. FALBE—Ruhrchemie AG Oberhausen, West Germany

R. SPECKS—Ruhrkohle AG Essen, West Germany

The demand for non-polluting or clean fuels to meet our ever-increasing requirements for energy is continuing to expand. This demand is superimposed on a world which over the years has preferentially exploited its clean energy reserves -- natural gas and high gravity low sulfur crude oils. As a result, there is continuing pressure on development of technologies to convert the remaining less desirable fuels -- coal, high sulfur crude, tar sand oil, etc., into clean non-polluting energy sources. This pressure is particularly strong in the United States and Japan.

In the United States and Japan, as well as other industrialized countries, the regulating agencies have established stringent air and water pollution standards. Perfecting new technologies that will meet these pollution requirements is a real challenge to the process developer.

Background. At the Texaco Montebello Research Laboratory in California, these challenges were recognized many years ago. Since the late 1940's this Laboratory has been carrying out research and development work on environmentally acceptable processes for converting heavy residual oils, tars, pitches, coal and coal liquifaction residues into synthesis gas. Synthesis gas, an approximately 50-50 mixture of hydrogen and carbon monoxide, can be efficiently converted to hydrogen, a much needed reactant in nearly all heavy fuel conversion technologies, using the long time commercially proven water gas shift reaction. Texaco's research efforts over the years have led to the development of two commercially viable processes - The Texaco Synthesis Gas Generation Process for gasification of fluids which are pumpable at temperatures

0-8412-0522-1/80/47-116-177$05.00
© 1980 American Chemical Society

as high as 600°F and The Texaco Coal Gasification
Process for gasification of solid carbonaceous materials which are fed to the gasifier as a slurry in
water or other carrier fluid.

Texaco Synthesis Gas Generation Process. For
many years, The Texaco Synthesis Gas Generation Process (1, 2, 3) has been available for licensing
throughout the world as an efficient technology for
converting high-sulfur residual petroleum fuels and
tars into synthesis gas. More than seventy-five
plants have been built in twenty-two countries since
the first units came on stream in 1955. Most of these
facilities have been associated with manufacture of
ammonia, methanol, and oxo-chemicals.

The synthesis gas generation process involves
reacting the residual fuel with a controlled amount
of high-purity oxygen and steam at pressures ranging
from 300-1200 psi with the net production of hydrogen
and carbon monoxide along with lesser amounts of
carbon dioxide and methane. The reactants are introduced through a special burner at the top of a refractory lined pressure vessel or generator, and the
desired autho-thermal non-catalytic reactions occur
rapidly at temperatures ranging from 2000-3000°F.
Small amounts of unconverted fuel or soot are recycled
to extinction. The approximate enthalpy changes at
60°F of the basic exothermic and endothermic chemical
reactions are shown in Figure 1. Sulfur present in
the fuel is converted to H_2S and small amounts of COS,
and organic nitrogen is reduced to elemental nitrogen
and ammonia. The hot exiting gases are cooled through
appropriate heat recovery equipment and treated as
necessary to produce the desired product, or the hot
gases are quenched in water in order to increase the
steam content of the gas so that it can be fed directly to a shift conversion reactor to convert the
carbon monoxide to hydrogen, if hydrogen is the desired end product. During shift conversion, the small
amount of COS in the gas is hydrolyzed to H_2S. Subsequently, the gas is treated for removal of CO_2, H_2S
and other undesirable impurities to produce a hydrogen
purity as dictated by the downstream hydrogen consuming process. A wide variety of commercially proven
gas treating technologies are available for the downstream gas purification.

Nearly thirty years ago it was recognized that
fuel for the process could just as well be coal, and
work on process development for production of synthesis gas from coal was initiated. At that time,

economic incentives for using coal in place of petroleum were not very strong, and the process development
proceeded on a low priority basis.

Texaco Coal Gasification Process.   In the late
sixties, the solids gasification process finally evolved to its present form and the energy crisis brought
on by the 1973 Arab oil embargo greatly accelerated
the development of the Texaco Coal Gasification Process.

The coal gasification process for hydrogen manufacture involves a slagging entrained down flow gasifier fed with oxygen and a concentrated slurry of
ground coal in water.   The same type of refractory
lined gasifier is employed as in the earlier oil
gasification process, except provision is made to
remove solidified slag through a water sealed lockhopper system.

Optionally, the gasifier may be fed with a slurry
of coal in oil and a controlled amount of reaction
temperature moderator, such as steam.   Facilities for
recycling unconverted coal are also provided.   A
schematic flow diagram of the process is shown in
Figure 2.

The process is capable of efficiently gasifying
a wide variety of caking and non-caking bituminous
and sub-bituminous coals, as well as petroleum coke.
Referring to Figure 2, raw coal is first fed to a
grinding section where the coal is ground either wet
or dry to a carefully controlled size distribution.
Control of the size distribution is important in order
to maximize the coal concentration in the resultant
slurry.   From the slurry preparation tank the coal
is then fed through a specially designed burner where
it is mixed with oxygen and additional moderator if
required.   Upon leaving the refractory lined gasifier
the gases are quenched in hot water and then the crude
raw synthesis gas is contacted with additional water
in a venturi or orifice type scrubber at gasifier
pressure to remove entrained particulates.   Water
removed from the scrubbing system and the quench
chamber is recovered through a settler where the
particulate matter is extracted and recycled to the
gasifier.   After water contacting, the particulate
loading in the raw gas is less than 1 mg/Nm^3.   The
water scrubbed gas is now ready for direct injection
into the shift conversion section.   Although a variety
of older catalysts are available for shifting sulfur
containing synthesis gas, the last ten years has seen
the development of high activity rugged cobalt moly

$$\sim \Delta H \; AT \; 60°F$$
$$BTU/LB\text{-}MOL$$

$$CH + H_2O \longrightarrow CO + 3/2H_2 \qquad +49,000$$

$$CH + 1/2O_2 \longrightarrow CO + 1/2H_2 \qquad -55,000$$

$$CH + 5/4O_2 \longrightarrow CO_2 + 1/2H_2O \qquad -229,000$$

$$H_2O + CO \longrightarrow CO_2 + H_2 \qquad -18,000$$

Figure 1.   Gasification reactions

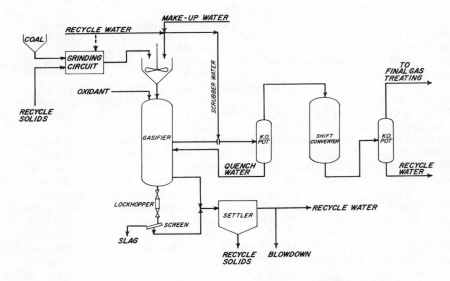

Figure 2.   Schematic flow diagram of Texaco coal gasification process for hydrogen manufacture

catalysts on an alumina base (4) which are especially
well suited to shift gases produced in the coal and
oil gasification processes. Relatively clean conden-
sate is recovered upon cooling the shifted gas which
is now ready for further gas treating. Since the
gasifier operates in a slagging mode, slag is continuously leaving the reaction zone of the gasifier with
the gas. Upon leaving the water in the quench section
of the gasifier relatively clean saturated gas is
separated. The slag accumulates in the lock hoppers
at the bottom of the quench chamber and is periodical-
ly withdrawn.

Slag removed from the dewatering screen below the
lock hoppers consists of fused particles generally
less than one-fourth to one-half inch in diameter.
This inert material containing less than 0.5 percent
carbon can be easily handled and removed to slag disposal sites. Typical slag removed from the pilot
unit at Montebello is shown in Figure 3.

In other applications where hydrogen is not the
ultimate product, substantially all the hot syn gas
from the gasifier may be directed through a heat
recovery system where the sensible heat of the gas is
used to generate steam for other process needs.

Environmental Aspects. Due to the high gasification temperature, by-product tars, phenols and
hydrocarbons heavier than methane are not produced.
Sulfur (H_2S) removed in the gas treating section is
routed to a Claus plant and tail gas treating unit
for conversion to elemental sulfur. Most of the
process water is recovered and recycled; however, a
small stream of blowdown is required to remove water
soluble inorganic material introduced with the coal.
In addition to the inorganic material, the withdrawn
water stream will contain small amounts of ammonia,
sulfide, formate, and cyanide. Tests have shown that
at detection limits ranging from 5 to 100 parts per
billion, no polynuclear aromatics on the EPA priority
pollutant list are present, and other organic compounds are only present at or below the detection
limits ranging from 5 to 30 parts per billion. Tests
on a typical blowdown water stream from the gasification of an Illinois coal are shown in Figure 4.

Leaching tests on the slag which must be removed
to a suitable disposal site have shown that the
material is essentially inert. Leaching tests on
slag from a typical Western coal are shown in Figure
5.

Figure 3.   Typical coal slag

pH	8.7
TOTAL ORGANIC CARBON	230 ppm
TOTAL INORGANIC CARBON	445 ppm
AMMONIA	1020 ppm
FORMATE	492 ppm
CHLORIDE	432 ppm
SULFIDE	264 ppm
SULFATE	166 ppm
CALCIUM	140 ppm
MAGNESIUM	100 ppm
SODIUM	80 ppm
THIOCYANATE	70 ppm
THIOSULFATE	69 ppm
FLUORIDE	39 ppm
CYANIDE	31 ppm
ALUMINUM	20 ppm
SILICON	5.0 ppm
IRON	3.7 ppm
C_6 + ORGANICS	
NAPHTHALENE	30 ppb
TOLUENE	20 ppb
BENZENE	10 ppb

Figure 4.   Typical water blowdown quality

Pilot Plant. At Texaco's Montebello Research
Laboratory there are two pilot gasifiers, Figure 6,
each capable of processing 15 to 20 tons per day of
coal. Tests on a wide range of coals have been conducted at pressures ranging from 300 to 1200 psi.
These pilot units along with the associated coal
grinding and slurry preparation equipment form the
basis for evaluating selected coals and provide design
information for a number of commercial projects that
are under way.

Composition of several of the feed stocks that
have been successfully gasified in the Montebello
pilot units is shown in Figure 7 and the composition
of the resulting synthesis gas is shown in Figure 8.

Demonstration Projects. Commercial and demonstration projects under way include a 150 T/D coal
gasification unit at the Ruhrchemie Chemical Plant
Complex in Oberhausen-Holten, West Germany (5) which
has been in operation for over a year. This research
and development demonstration plant is jointly funded
by Ruhrchemie A.G., Ruhrkohle A.G. and the Government
of the Federal Republic of Germany.

The project was designed to demonstrate that the
Texaco Coal Gasification Process can be employed to
generate synthesis gas which can be used as a feed
stock for the chemical industry. The design conditions of the Oberhausen-Holten plant are given in
Figure 9. The design basis includes recycle of unconverted carbon and is based upon test runs made on
the Montebello pilot unit on a prototype coal from
the Ruhr region of Germany. The ratio of hydrogen to
carbon monoxide can of course be increased by partial
or complete shift to provide synthesis gas or hydrogen
for a wide variety of chemical synthesis processes.

A process flow diagram of the demonstration plant
is shown in Figure 10. The design of the Oberhausen-
Holten plant differs in two sections from that of the
Montebello pilot unit.

The coal slurry is prepared in either of two wet
grinding systems, both of which have demonstrated
efficient operation and ability to prepare coal
slurries of the desired concentration. Secondly, the
water quench system at the exit of the gasifier has
been replaced with a radiation cooler.

Slurry is stored in an agitated surge tank from
which it is fed to the gasifier by a variable speed
pump. The metered coal slurry enters the top of the
gasifier along with a controlled flow of oxygen. Hot
gases and slag exit the gasifier and enter a down-flow

pH	6.85
CONDUCTIVITY, micromhos	350
TOTAL DISSOLVED SOLIDS	260 ppm
SUSPENDED SOLIDS	12 ppm
CHEMICAL OXYGEN DEMAND	52 ppm
TOTAL ORGANIC CARBON	16 ppm
SULFATE	104 ppm
CALCIUM	75 ppm
CHLORIDE	12 ppm
SODIUM	8.5 ppm
POTASSIUM	325 ppb
IRON	224 ppb
MAGNESIUM	167 ppb
ALUMINUM	110 ppb
MANGANESE	97 ppb
NICKEL	30 ppb
LITHIUM	25 ppb
ZINC	23 ppb
LEAD	13 ppb
CHROMIUM	12 ppb
MERCURY	4 ppb

Figure 5. Typical slag leaching tests: 500 g ground slag and 2000 g distilled water agitated for 24 hr. Filtered leachate properties.

Figure 6.   Coal gasification pilot units

WEIGHT PER CENT	WESTERN BITUMINOUS COAL	DELAYED PETROLEUM COKE	EASTERN BITUMINOUS COAL	SRC II VAC.TOWER RESIDUE
C	74.5	89.5	67.6	63.7
H	5.3	3.7	5.2	3.6
S	0.5	1.4	3.3	2.9
N	1.0	2.7	1.0	1.2
O	11.5	2.4	11.1	0.8
ASH	7.2	0.3	11.8	27.8
HEAT OF COMBUSTION, BTU/LB DRY	12,500	14,950	12,270	11,250

Figure 7.   Properties of typical feedstocks

FEEDSTOCK	WESTERN BITUMINOUS COAL	DELAYED PETROLEUM COKE	EASTERN BITUMINOUS COAL	SRC II VAC.TOWER RESIDUE
OXIDANT	OXYGEN	OXYGEN	OXYGEN	OXYGEN
SLURRY MEDIUM OR MODERATOR	WATER	WATER	WATER	STEAM
PRODUCT GAS COMPOSITION, VOLUME PER CENT DRY BASIS				
H_2	35.8	34.5	35.8	34.0
CO	50.7	45.2	44.6	59.5
CO_2	13.1	19.4	18.0	5.3
N_2 -A	0.2	0.7	0.5	0.6
CH_4	0.1	—	—	0.1
H_2S	0.1	0.2	1.0	0.5
COS	—	—	0.1	—

Figure 8.   Typical gasification performance summaries

PRESSURE	40 bar
TEMPERATURE	1350-1450 °C
SOLIDS CONTENT	55-65 %
GAS ANALYSIS H_2	30-40 %
CO	45-55 %
CO_2	15-20 %
CH_4	< 1 %
CARBON CONVERSION	> 98 %
COLD GAS EFFICIENCY	> 70 %
COAL FEED	6 t/hr
GAS PRODUCTION	12000 m³/hr

Figure 9.   General design data for Oberhausen–Holten coal gasification demonstration plant

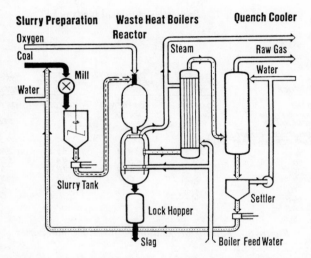

Figure 10.   Process flow diagram of the Oberhausen–Holten coal gasification demonstration plant

water walled radiation section where they are substantially cooled and the slag is solidified, simultaneously generating high pressure steam in the water
walled section. Slag is removed through a water seal
and lockhopper system for disposal. Hot gases and
entrained fly ash exit above the water seal into a
fire tubed boiler where additional steam is generated,
the gas is cooled to approximately 280°C and finally
the cooled gases are quenched and water scrubbed. The
clean gas is then sent to a boiler, or treated in an
acid gas removal system for use in the chemical plant.
A photograph of the demonstration plant is shown in
Figure 11.

Operation of components of the coal gasification
plant commenced at the end of January 1978. Between
April 1978 and March 1979 more than 50 test runs have
been conducted covering more than 2000 hours. Several
test runs demonstrated continuous operation for
periods in excess of 300 hours. Approximately 8000
metric tons of different Ruhr area coals of varying
volatility have been gasified at pressures ranging
from 300 to 600 psig. Without unconverted coal
recycle carbon conversions up to 96% and cold gas
efficiency as high as 72% have been achieved. Steam
generated in the waste heat boiler corresponds to
about 20% of the heat of combustion of the coal.

Further development and process optimization is
planned during the second year of the demonstration
program. Included in the program will be evaluation
of alternate components, testing alternate concepts
for heat recovery, and gasifying a wider range of
coals.

The total program is planned to provide sufficient information to confidently design large scale
second generation coal gasification plants (100,000
Nm^3/hr of synthesis gas) using the Texaco process.

In addition to the demonstration plant in West
Germany, the Tennessee Valley Authority is converting
a natural gas fueled ammonia plant in Muscle Shoals
to coal by installing a 150 T/D Texaco coal gasifier
and coal slurry preparation system. This plant
scheduled to gasify Illinois coal is presently in the
detailed engineering design stage.

W. R. Grace and Ebasco, with funding from the
Department of Energy, are also in the engineering
design phase of a large grass roots 1200 T/D ammonia
from coal plant employing Texaco gasifiers operating
at 1200 psi on Eastern U.S. coal.

Figure 11.   Oberhausen–Holten coal gasification demonstration plant

In addition to the above projects, Texaco and
Southern California Edison have jointly announced
their intention to obtain partial support from the
Electric Power Research Institute and others to construct a 1000-T/D coal gasification combined cycle
demonstration plant near Barstow in the high desert
northeast of Los Angeles (6, 7). Fuel gas from the
coal gasifier will be routed to the combined cycle
power generation facility which will produce 90
megawatts of electrical power through gas and steam
turbines. The gas from the coal gasifier can also be
used to fire a steam boiler in an existing power generating facility which produces 65 megawatts of electricity. Preliminary engineering on this project is
under way.

Economics. At the present time, process economics are difficult to evaluate. Costs of hydrogen,
low BTU fuel gas, or electric power are dependent on
plant size, coal cost, plant location, product purity,
and other environmental requirements that must be met.
Using mid-1979 dollars, we have estimated 99% purity,
1000 psig hydrogen from a 50MM to 100MM SCF/D plant to
cost between $1.25 to $1.60 per 1000 SCF ($4-$5 per
MM BTU) before taxes and profit.

Fluor Engineers and Constructors Inc., under
contract with EPRI has completed a rather detailed
study (8) of the cost of producing power via Texaco
gasifiers and advanced design gas turbines and steam
turbines. They conclude investment costs (mid-1976
dollars) to be $815 million for a plant gasifying
9600 T/D of moisture-free Illinois coal and producing
1150 megawatts of net power.

Obviously, the Texaco entrained bed slagging
gasifier has many varied applications, only a few of
which have been discussed today. It is anticipated
that large coal conversion plants will be operational
in many parts of the world by the 1990's, making a
significant contribution to the total clean energy
supply of the industrialized world.

LITERATURE CITED

1. Marion, C.P. and Slater, W.L. "Manufacture of Tonnage Hydrogen by Partial Combustion - The Texaco Process"; Sixth World Petroleum Congress, Frankfurt, Germany, June 1963.

2. Child, E.T. "Texaco Heavy Oil Gasification"; The University of Pittsburgh School of Engineering Symposium on Coal Gasification and Liquifaction, Pittsburgh, Penn. August 6-8, 1974.

3. Robin, A.M. and Schlinger, W.G. "Gasification of Coal Liquifaction Residues"; The 13th Intersociety Energy Conversion Engineering Conference, San Diego, California, August 1978.

4. Auer, W.; Lorenz, E.; Grundler, K.; "A New Catalyst for the CO Shift Conversion of Sulfur-Containing Gases"; 68th National Meeting of the AIChE, Houston, Texas, February 28-March 4, 1971.

5. Specks, R.; Langhoff, J.; Cornils, B.; "The Development of a Pulverized Coal Gasifier in the Prototype Phase Using the Texaco System"; Coal Refining Symposium, Edmonton, Canada, April 20-21 1978.

6. Electric Power Research Institute Final Report AF-880; "Preliminary Design Study for an Integrated Coal Gasification Combined Cycle Power Plant"; EPRI, Palo Alto, California, August 1978.

7. Gluckman, M.J.; Holt, N.A.; Alpert, S.B.; Spencer, D.F.; "The Near Term Potential for Gasification - Combined Cycle Electric Power Generation"; Energy Technology VI Conference, Washington, D.C. February 26-March 1, 1979.

8. Electric Power Research Institute Final Report AF-642; "Economic Studies of Coal Gasification Combined Cycle Systems for Electric Power Generation"; EPRI, Palo Alto, California, January 1978.

RECEIVED September 17, 1979.

Production and Application of Electrolytic Hydrogen: Present and Future

L. J. NUTTALL

Direct Energy Conversion Programs, General Electric Company,
50 Fordham Road, Wilmington, MA 01887

Water electrolysis is an old technology for generating hydrogen. With the exception of a few special applications, however, it has not been economically competitive with steam reforming of natural gas. In 1975 there were approximately 1.8×10^{12} cu ft of hydrogen produced in the United States of which less than 1% was produced by water electrolysis. The ratio on a world-wide basis was undoubtedly higher since the primary applications for electrolysis have been in areas where low cost hydroelectric power is available, i.e., Norway and Egypt. However, the percentage is still very small.

The role of electrolysis in the production of hydrogen is certain to increase in future years, but the extent to which it does depends on a number of factors, including:

- Price and availability of natural gas relative to electric power
- Rate of development of non-fossil fuel energy sources
- Improvements in electrolyzer technology

One of the most significant current activities relating to this last item is a development program underway at General Electric to develop an advanced electrolysis technology for large scale commercial hydrogen generation using the solid polymer electrolyte type of cell originally developed for aerospace fuel cell and electrolysis applications.

General Economics

At the present time hydrogen can be produced in a large, e.g., 100 MCF/D (M = million, 10^6; K = thousand, 10^3) steam reforming plant for a cost in the range of $6/MBTU ([1]). This compares with an estimated cost for electrolytic hydrogen of $15 to $20/MBTU, using current conventional electrolyzers, depending on the local cost of electrical power. The advanced solid polymer electrolyte electrolysis technology has the potential to reduce the cost for electrolytic hydrogen under comparable conditions to the range of $9-$13/MBTU.

Figure 1 shows the calculated cost of hydrogen as a function of the electrolyzer plant cost, in dollars per KW (calculated from the higher heating value of the hydrogen produced — i.e., 10.7 SCFH 1 KW) of output and the overall system efficiency (ratio of KW output to the electrical power input), for an electric power cost of

0-8412-0522-1/80/47-116-191$05.50
© 1980 American Chemical Society

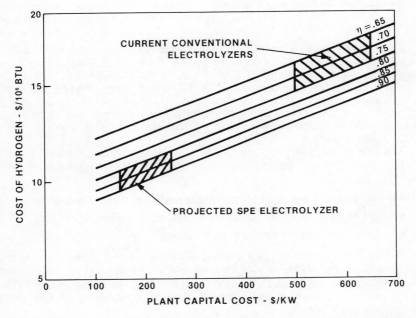

Figure 1.   Estimated hydrogen production cost from a large water electrolysis
plant. Electric power cost = 2.5¢/kWh; 90% duty cycle.

25 mils/KWH. The plant cost in this calculation is the total investment cost, including on-site, off-site and indirect capital costs associated with the plant construction. The approximate range of characteristics for the conventional electrolyzer as well as those projected for the SPE units are noted on this curve.

Even the costs indicated for the advanced SPE electrolyzer are not competitive with steam reforming of natural gas at current prices in large plants. However, for smaller plants, the cost for generating hydrogen in a reformer increases at a faster rate than that for an electrolysis unit. Therefore, it appears, as will be shown later, that the SPE electrolyzer will be competitive for plants in the range of 400-500 K SCF/D and smaller. For larger plants, electrolysis will probably be used only in selected applications where lower cost electrical power can be made available. The estimated cost of electrolytic hydrogen as a function of electric power cost for an 85% plant efficiency is shown in Figure 2. One example would be in locations where low cost hydro-electric power is available. There are a number of locations around the world where it has been estimated that electric power can be provided at a cost between 5 and 10 mils/KWH. In such cases, an electrolysis plant can be more economical than gas reformation, regardless of the output.

Another possibility for the future, where very large quantities of hydrogen are involved, is to consider a dedicated nuclear plant integrated with a water electrolyzer. This is analogous to the type of facility being investigated in conjunction with the various thermochemical water splitting approaches for hydrogen generation. Two recent studies (2, 3) have indicated that, with the economics that can be realized through integrating of the power generation cycle with the electrolysis cycle, the cost for electrolytic hydrogen can be reduced to around $6/MBTU (including credit for the by-product oxygen), which is competitive with steam reforming of natural gas and does not rely on a dwindling fossil fuel resource.

These studies also indicate an overall efficiency (ratio of higher heating value of hydrogen output to the nuclear heat input) in the range of 40 to 43%, which is comparable to that presently estimated for the most promising thermochemical water splitting cycles.

Development of other non-fossil fuel energy sources, such as geothermal, solar, wind and ocean thermal systems will undoubtedly broaden the scope of application for electrolytic hydrogen — perhaps not because they represent a source of low cost electrical power, but because it represents the most attractive means for providing a transportable fuel from these renewable resources.

Solid Polymer Electrolyte Electrolysis Technology

The solid polymer electrolyte is a solid plastic material which has ion exchange characteristics that make it highly conductive to hydrogen ions. The particular material that is used for the current electrolysis cells is an analogue of TFE teflon to which sulfonic acid groups have been linked. This plastic sheet is the only electrolyte required, there are no free acidic or caustic liquids, and the only liquid used in the system is distilled water.

A typical cell employs a sheet of the polymeric material approximately 10 mils thick. A thin catalyst film is pressed on each face of this sheet to form the anode and cathode electrodes. Since the electrolyte is solid, the electrodes do not have to perform any structural or containment functions. Consequently, they are very simply designed for the sole function of providing sufficient catalytic activity to achieve desired performance levels.

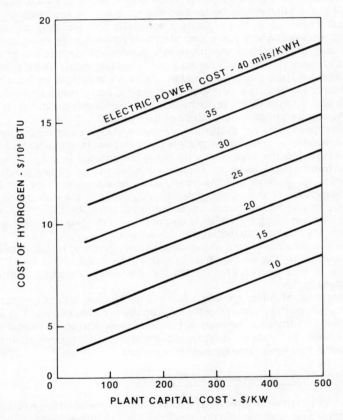

Figure 2.   Estimated hydrogen cost from electrolysis plant. System efficiency =
85%; duty cycle = 0.9.

Water is supplied to the oxygen evolution electrode (anode) where it is electro-chemically decomposed to provide oxygen, hydrogen ions and electrons. The hydrogen ions move through the sheet to the hydrogen evolving electrode (cathode) while the electrons pass through the external circuit. At the hydrogen electrode, the hydrogen ions and electrons recombine electrochemically to produce hydrogen gas.

An excess of water is supplied to the cell and recirculated to remove waste heat.

This technology was originally developed as a fuel cell power source for the Gemini spacecraft. However, it was adapted for use as an electrolyzer beginning in the early years of this decade. Typical current applications include a spacecraft regenerative life support system (Figure 3) and an oxygen generation system for nuclear submarines, the electrolysis stack for which is shown in Figure 4.

A small commercial unit for generation of pure hydrogen for gas chromato-graphs and other laboratory uses has been on the market since 1973.

In 1976, a program was initiated to develop a large-scale electrolysis design that will be suitable for bulk hydrogen generation for industrial and utility applications. This program is sponsored jointly by the U.S. Department of Energy, some of the electric and gas utilities, and the General Electric Company.

The initial phase of this program was a design study of a 58 MW (625,00 SCFH of H_2) system which would be suitable for a number of potential applications including energy storage, generating hydrogen for chemical and industrial feedstock or possibly as a supplement to natural gas in certain areas.

Figure 5 is a model of a typical energy storage plant based on the results of the study. The 58 MW water electrolysis system is shown cut away in the foreground, with power conditioning housed in the rear of the building. To the right are the metal hydride hydrogen storage cylinders. A typical module from a 26 MW air-breathing fuel cell installation for conversion back to electricity is shown to the rear of the building.

On the basis of the study results, the goals for the development program were established as follows:

Overall System Efficiency	85-90%
System Capital Cost (Battery Limits)	$100/KW
Scale-Up	5 MW Demo System

Technology Development

The efficiency goal for this program requires a low cell operating voltage, and the cost goal requires a high operating current density as well as a low manufacturing cost. The key advantage of the SPE cells is the superior performance which makes the high current density possible at a low cell voltage. Figure 6 shows the current SPE electrolyzer performance compared with that of conventional alkaline units and also shows the improved performance which is expected to be achieved by the completion of this program. The design current density for most of the applications of this technology is around 1000 amps/ft^2.

The other objective of the program is to reduce the manufacturing cost of the SPE cells for commercial application by a factor of about 14:1 from the cost of those used in the space and submarine systems.

To date, the technology development program has resulted in significant progress toward these goals. Compared with a 1975 baseline technology of $202/KW,

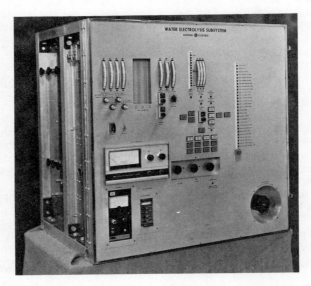

Figure 3. Oxygen life support system for manned spacecraft and space station applications

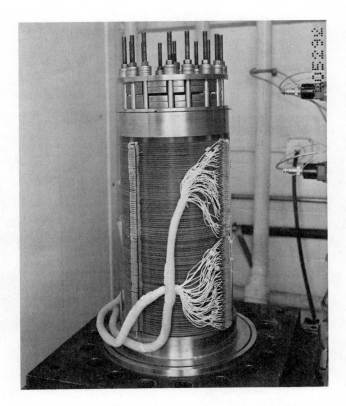

Figure 4. A 100-cell electrolysis module for submarine life support

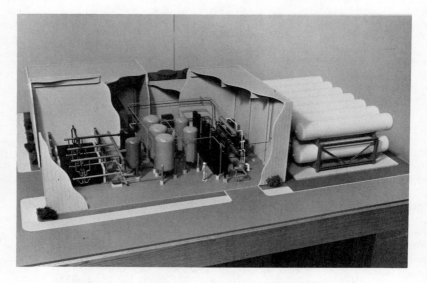

Figure 5.   A 58-MW–SPE water electrolysis plant

projected cost of electrolyzer production based on currently identified materials and techniques is about $50/KW or 80% of the cost reduction to meet the 1983 goal. The primary cost reductions have been in the areas of current collectors and catalytic electrodes.

Molded carbon current collectors are a major result of the technology development program. The function of the current collector is to:

- Provide the flow fields for the water and oxygen on the anode (oxygen) side and for the water and hydrogen on the cathode (hydrogen) side
- Separate the oxygen and hydrogen sides
- Provides for the conduction of electricity from one cell to the next

The collector is molded from a mixture of carbon and phenolic resin which incorporates an in-situ formed titanium foil shield on the anode side to prevent corrosion. Small laboratory-sized collectors have accumulated over 12,000 hours of operational evaluation to date. Figure 7 shows large-sized molded collectors with $2\frac{1}{2}$ ft^2 active area.

A major advantage of the molded collector is the elimination of costly silicone rubber gaskets. The SPE itself acts as a gasket between the sealing surfaces of the collector. Gasketless sealing has been demonstrated up to 500 psi on laboratory-sized cells, and, in a previous prototype hardware program, up to 400 psi on a 120 cell stack.

Catalytic electrode development has involved the identification of a ternary oxygen evolution catalyst which offer both reduced cost and improved performance in comparison to the current state-of-the-art for aerospace systems.

Figure 8 shows the cost reduction and performance improvements of these new catalysts demonstrated in operational cells. Effort is continuing in both data base testing of currently identified catalyst systems, and identification of additional catalyst systems.

The catalyst identifications shown on this figure are merely G.E. designations. The catalyst compositions are considered proprietary.

The results of reductions in catalyst loading to effect a cost improvement with minimum impact on cell performance have also been encouraging. Techniques for reducing both anode and cathode loadings up to 93% have been identified and shown feasible.

High temperature operation (up to 300°F) offers advantages in operating efficiency and accounts for about one half of the improvement needed to meet the goal performance shown on Figure 6. Over 5000 hours have been demonstrated to date at 300°F using carbon collector components and Nafion ® 120.

Alternative solid polymer electrolytes are also being evaluated to assess advantages in cost and performance. The Nafion ® 120 is an extremely stable (and thus long-lived) material under water electrolysis operating conditions. Any alternative SPE considered viable must have equivalent life stability. A radiation-grafted trifluorostyrene was extensively evaluated, but demonstrated insufficient operational stability to be considered as a viable alternative. The search for other alternatives is continuing.

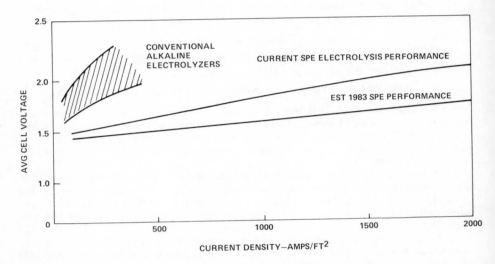

Figure 6. *Comparative water electrolysis performance*

Figure 7. *Molded current collector 2.5 ft² active area*

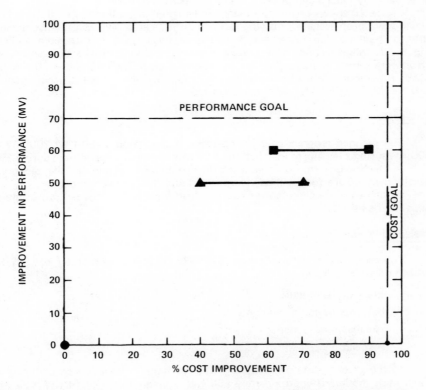

Figure 8. Anode catalyst development: (●), E = 50; (■), E = 100; (▲), WE = 3.

Scale-up

The cell scale-up program is proceeding in a parallel with the technology development program. The cell scale-up is planned in two steps: development of a 2½ ft^2 active area cell, which started in 1977, followed by a 10 ft^2 cell development, which will start this year.

Figure 9 shows one of the 2½ ft^2 electrolyte and electrode assemblies which, in conjunction with a molded carbon current collector (shown in Figure 7) and suitable anode and cathode supports, form a bipolar cell. The cells are assembled between pneumatically loaded end plates to form an electrolysis module.

Figure 10 shows one of the many single and multi-cell modules of 2½ ft^2 cells which have been tested during the last 6 months. Large cell performance, comparable to "baseline" laboratory cell performance, has been demonstrated as shown in Figure 11. Initial testing concentrated on sealing and fluid distribution, and present efforts are aimed at increasing the number of cells in the module. A 50 KW (500 SCFH H$_2$) module will soon be on test and assembly of a system to accommodate a 200 KW (2100 SCFH H$_2$) module has started.

Future Plans

The planned program calls for operational evaluation of 50 KW, 200 KW and 500 KW systems leading to installation of a 5 MW demonstration system during 1983.

As technology improvements are identified and proven by the technology development effort, they will be incorporated into the hardware program.

As hardware development progresses, field installation of small prototype systems will be initiated.

Projected Production Cost

The calculated system cost from the 1975 58 MW system study was $82/KW, broken down as follows:

Electrolysis Module	$14/KW
Power conversion and control	43
Ancillary components	16
Installation	9

At the present state of technology the estimated production cost of a 58 MW electrolysis module is approximately $50/KW compared to the $14/KW projection.

The impact of this difference in electrolyzer cost on the system cost is shown in Figure 12. The program goal of $100/KW allowed for some difficulties in achieving all of the calculated cost bogies, and, with today's technology, it appears that the capital cost for a 58 KW system could be approximately $118/KW compared with the $100/KW goal.

For smaller installations the cost per KW for the electrolysis system will increase. Figure 12 also shows the estimated cost as a function of the capacity of the system for lower capacity units.

Figure 9.   A 2.5 ft² active area membrane and electrode assembly

Figure 10.   A 2.5 ft² electrolyzer cell module

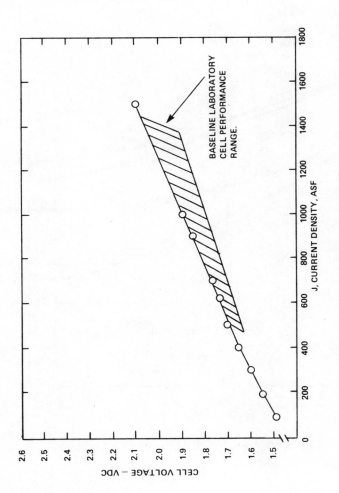

Figure 11.   A 2.5 ft² cell test performance. Cell no. = DOE 12; operating temperature = 108°F; generating pressure = 50 psig.

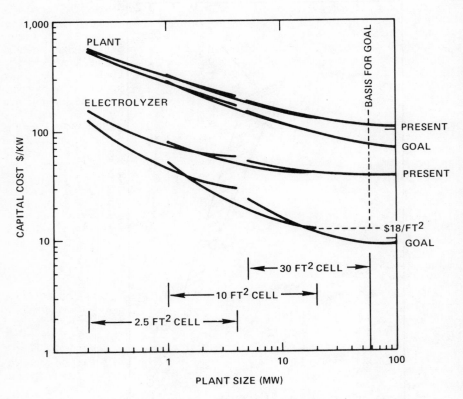

Figure 12.   Installed cost vs. plant size

Future Hydrogen Production Costs

As mentioned previously, one of the factors which will influence the future role of electrolytic hydrogen is the relative rates of increased costs between electrical power and natural gas. A projection that was made by one General Electric component is shown on Figure 13, indicating that by 1985, they expect natural gas to start increasing in price at a faster rate than electrical power. Assumed inflation rates of 6.5% through 1985 and 5% from 1985 to 2000 are factored into these curves. On the basis of this projection, the cost for producing hydrogen by SPE water electrolysis and by steam reforming of natural gas was calculated for several different sizes of plants. Figure 14 shows the results for a large plant of 100 MSCF/D. In this case, electrolytic hydrogen, using commercial electrical power, will not be competitive with that from natural gas until after the year 2000. However, a dedicated nuclear/electrolysis plant, which could be available in the 1990's, has the potential to produce hydrogen at a significantly lower cost than either of the above systems. In fact, the cost for the hydrogen from this system appears to be competitive with that shown in Figure 13 for natural gas by the year 2000, so it should also be considered as a possible source of SNG for fuel applications in future years.

Figure 15 shows the results for two smaller plants, which indicate that the electrolysis system begins to show an economic advantage for plant capacities below about 400-500K SCF/D. This conclusion has also been confirmed in a study performed by Hittman Associates for the Department of Energy (4). One of the figures from that study is reproduced in Figure 16 showing their estimate of the cost of hydrogen from natural gas reforming as compared with that from a SPE electrolyzer. This curve also shows the price ranges for truck delivered hydrogen as a function of use rate, indicating that there will probably also be an advantage to the small users who now purchase truck delivered hydrogen to consider installing an on-site electrolysis system.

Conclusion

It is apparent from this analysis that the large scale use of water electrolysis for hydrogen production is a number of years away. During the 1980's there may be a number of large electrolysis plants installed in selected applications where low cost hydro-electric power is available, or where some utilities may want to use hydrogen to assist them with load management. However, the general applications during this period will undoubtedly be for small industrial applications where an on-site electrolysis system works out to be more economical than purchased gas or an on-site reformer.

In later years the use of electrolysis can be expected to expand as a greater portion of the world's energy begins to be derived from non-fossil fuel sources. Electrolysis remains the most promising means to produce hydrogen from nuclear or the various forms of solar energy for use either as a chemical feedstock or a synthetic fuel.

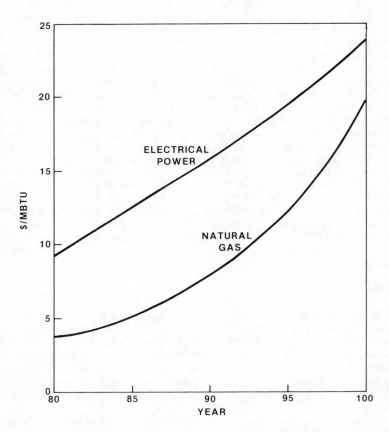

Figure 13.   Projected cost for electrical power and natural gas

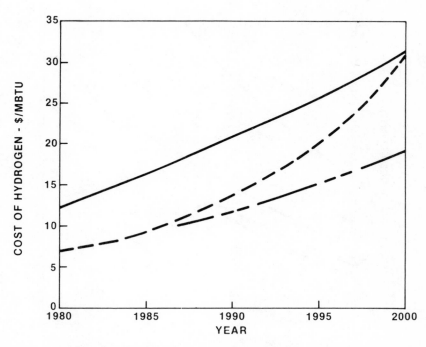

Figure 14.   Projected hydrogen cost of a large plant (100,000,000 scfd): (——),
water electrolysis; (– – –), steam reforming of natural gas; (— — —) dedicated
nuclear–water electrolysis.

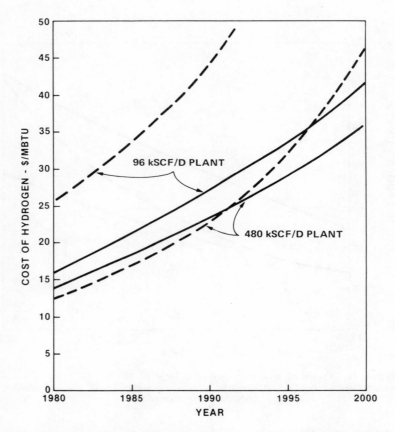

*Figure 15. Projected hydrogen cost from smaller plants: (——), water electroly-
sis; (– – –), steam reforming of natural gas.*

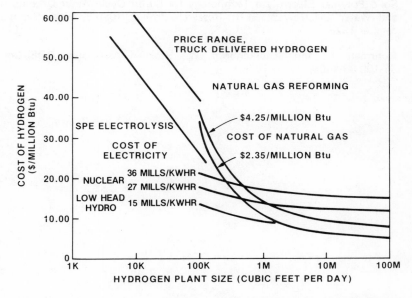

Figure 16. Comparative costs of producing hydrogen: advanced electrolysis vs. natural gas reforming

References

(1) Corneil, H.G., Heinzelman, F.J. and Nicholson, E.W.S., Production Economics for Hydrogen, Ammonia and Methanol During the 1980-2000 Period, Exxon Corporation, Report No. BHL-50663, Contract No. 368150-5.

(2) Escher, W.J.D., Donakowski, T.D. and Foh, S.E., Dedicated Nuclear Facilities for Electrolytic Hydrogen Production, Institute of Gas Technology, Presented at the Hydrogen for Energy Distribution Symposium, July 24-28, 1974.

(3) Russell, J.H., Sedlak, Dr. J.M., General Electric Company, Direct Energy Conversion Programs, Economic Comparison of Hydrogen Production Using Solid Polymer Electrolyte Technology for Sulfur Cycle Water Decomposition and Water Electrolysis, EPRI Research Project 1086-3, Final Report, December 1978.

(4) Dohrman, D. and Scherkenbech, W., Hittman Associates, DOE Contract 31-109-38-4409, March 1978.

RECEIVED July 13, 1979.

COMMERCIAL DISTRIBUTION AND SAFETY

Safe Handling of Hydrogen

CLYDE McKINLEY

Air Products & Chemicals, Inc., P.O. Box 538, Allentown, PA 18105

Fuel of the Future. Hydrogen has long been considered the
fuel of the future because of its clean combustion characteristics, its adaptability in simple fuel cell use, its potential
for distribution by a grid similar to that used for natural and
manufactured gases, and the limitless supply available using
electrical energy via nuclear energy. Today, economics limit
its use to applications such as in the refining, petrochemical,
metallurgical industries, as well as in ammonia and methanol
where it has greater value than it would have as a competitive
fuel. Its fuel uses are now extremely limited. This balance
may change relatively rapidly with changes in the world's energy
situation.

Hydrogen Can Be Handled Safely. The safe handling of
hydrogen in its production, storage, and use has now been demonstrated in many large scale industrial, military, and space
applications. The techology has now been developed. Scientific
foundations have been laid and engineering design and operating
practices demonstrated to fully identify all hazards involved in
handling hydrogen and to provide the base for the practice of
all needed safety precautions. It is the purpose of this paper
to highlight the principal hazards and the information needed to
minimize the risks with each hazard.

There remains a concern among those now knowledgeable with
hydrogen in considering its widespread use, a concern probably
related to remembering the Hindenberg at Lakehurst, and associ-
ating hydrogen with the H-bomb. Knowledge now allows us to
better understand hydrogen and to safely handle it in any situation where its use is justified.

Hydrogen Handling Hazards. The principal hazards associated
with handling hydrogen as a gas or a liquid or a slush (mixture
of solid and liquid hydrogen) are related to

0-8412-0522-1/80/47-116-215$05.00
© 1980 American Chemical Society

° Flammability of hydrogen
° Expansion of liquid hydrogen to gaseous hydrogen in confined
  spaces
° Low temperatures of liquid or slush hydrogen adversely
  affecting the containment materials subjected to the low
  temperatures
° Cold "burns"
° Suffocation and breathing atmospheres

Each of the foregoing hazards is discussed in the following
sections. Table 1 provides certain specific properties of
hydrogen useful in examining the hazards discussed in the following sections (1).

Flammability of Hydrogen

The Fuel-Oxidant-Ignition Hazard. Avoidance of the flammability hazard requires understanding of hydrogen's peculiar
physical-chemical properties as related to the classical flammability problem. There must be a fuel, in our case hydrogen;
there must be an oxidant available in flammable proportions,
typically oxygen from air; and there must be a means of ignition.
With an understanding of hydrogen's properties, systems for handling hydrogen may be designed to allow safe operation. Hazard
reviews, establishment of safe operating practices, and proper
training make possible safe operation of properly designed
systems.

Hydrogen-Nitrogen-Oxygen Mixtures (2-5). The lower limit
of hydrogen concentration for flame propagation at ambient
temperature and atmospheric pressure in air and in oxygen is
conservatively listed as 4.0 volume percent. A higher concentration is required for horizontal or downward propagation of
the flame, as we will note in a following section. The 4.0
percent value (Point A) is low and therefore may be considered
conservative in hazard evaluation. The upper limit of hydrogen
concentration in air is 75 percent (Point B) and in oxygen is 95
percent (Point C). These values are shown in Figure 1. The
shaded area inside the triangle bounded by the lower limit for
oxygen, the lower limit for hydrogen, and the H_2-O_2 side of the
triangle shows all possible flammable mixtures of H_2-N_2-O_2. At
the lower limit boundaries of the triangle the gas mixtures upon
combustion release barely sufficient heat to propagate a flame.
Gas mixtures which lie within the triangle, away from its lower
limits boundaries, release more heat upon combustion, burn with
higher velocities and may detonate. The detonable limits for
hydrogen in air are noted; the lower and upper limits, 18 and 59
percent, points D and E. The detonable limits for hydrogen in
oxygen are 15 and 90 percent, points F and G. The central
triangle shows the concentration of detonable mixtures.
Hydrogen-air mixtures in the detonable region may burn at velo-

Table I. Properties of Hydrogen (Normal)

Molecular weight	2.0159 (9)
Triple-point temperature	13.957 K (9)
Triple-point pressure	0.0711 atm (9)
Triple-point liquid density	38.3 mol/L (10)
Triple-point solid density	43.01 mol/L (10)
Triple-point vapor density	0.0644 mol/L (10)
Normal boiling point	20.39 K (1)
Normal boiling-point liquid density	35.2 mol/L (9)
Normal boiling-point vapor density	0.6604 mol/L (9)
Critical temperature	33.19 K (9)
Critical pressure	12.98 atm (9)
Critical volume	14.94 mol/L (9)
Latent heat of fusion at triple point (p-H_2)	28.08 cal/mol (9)
Latent heat of vaporization at nbp	214.5 cal/mol (9)
Heat of combustion, gross	68317 cal/mol (11)
Heat of combustion, net	57798 cal/mol (11)
Limits of flammability in air	4.0 to 75.0 vol % (3)
Limits of detonability in air	18 to 59 vol % (3)
Burning velocity in air	up to 2.6 m/sec (6)
Burning velocity in oxygen	up to 8.9 m/sec (6)
Limits of flammability in oxygen	4.0 to 95 vol %
Limits of detonability in oxygen	15 to 90 vol %
Detonation velocities of H_2-O_2 mixtures	
15% H_2 in oxygen	1400 m/sec (6)
90% H_2 in oxygen	3600 m/sec (6)
Spontaneous ignition temperature	520° to 580°C (4)

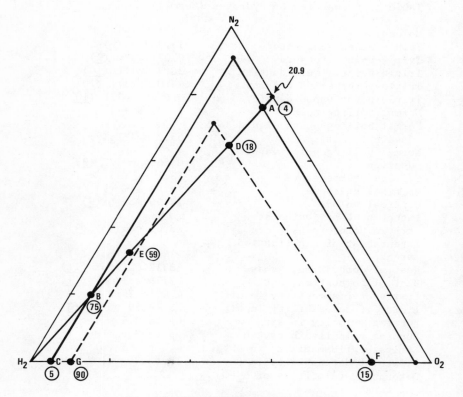

Figure 1.   H₂–O₂–N₂ flammability triangle

cities in the range from less than one meter per second up to
8.9 meters per second. With strong ignition or with confinement
these mixtures may pass over from relatively slow burning velocities to detonation velocities up to 3600 meters per second.

Diluents and Inert Materials. Inert diluent gases, when
added to flammable mixtures of hydrogen with air and/or oxygen,
will cause a reduction in flame temperature and burning velocity
until flame propagation is no longer sustained. The hydrogen in
such non-flammable mixtures would be consumed if the mixture
were preheated or passed over a suitable catalyst.

Flammable Limits, A Function of Direction of Flame Travel
(3). Hydrogen flammable limits are peculiar. The heat of
combustion of a lower limit mixture of most gases will result in
a flame temperature much higher than the ignition temperature of
the mixture. The lower limit of hydrogen in air, 4.0 volume
percent for upward propagation, produces a calculated average
flame temperature of less than 350°C, whereas the ignition temperature of hydrogen in air is 585°C. This very low average
temperature can be understood from observations that the rising
flame in a limit mixture rises as luminous balls, consuming only
a part of the hydrogen. Fresh hydrogen diffuses into the burning
ball yielding a higher effective concentration of hydrogen than
was present initially. It has been observed that not all of the
hydrogen is consumed in an upward propagating flame in a 2 inch
tube until 10 percent hydrogen was present. Similar experiments
with horizontal tubes result in a lower flammability limit of
about 6.5 percent hydrogen; downward flame propagation requires
about 9.0 percent hydrogen in air. The upper limit is about 75
percent for all propagation directions.

Effect of Pressure and Temperature on Flammable Limits (3).
Pressure has little effect upon the lower limit of hydrogen in
oxygen, for downward flame propagation, with pressures ranging
from atmospheric pressure up to 122 atmospheres. Over this
range the lower flammability limit ranges from 8 to 10 percent
as determined by several different experimentalists. Temperature of the preignition mixture has a marked effect, as shown in
Figure 2. These particular data show 9.5 and 71.5 for the two
limits at ambient temperature. These limits broaden linearly
with rising temperature to 6.3 and 81.5 percent hydrogen at
400°C. The possibility for elevated temperatures in a preheated
hydrogen-air mixture must be remembered in evaluating a potentially hazardous situation. Preheating may cause an otherwise
non-flammable safe mixture to become flammable.

Ignition Energy (6). The ignition energy must be sufficient
to establish a flame of critical minimum size. If the ignition
energy input is less than this minimum the initially established

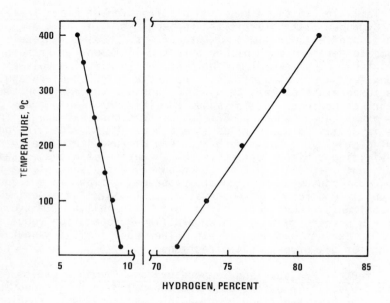

Figure 2.   Hydrogen flammability limits vs. temperature

flame will be quenched by ambient unburned gas or other system components. Minimum ignition energy is very much a characteristic of the system. With spark ignition, the minimum ignition energy for 30 percent H_2 in air 0.015 millijoules. For a similar experimental system the minimum ignition energy for the C_1 through C_6 paraffinic hydrocarbons, stoichiometrically mixed with air, is about 0.25 millijoules (over an order of magnitude greater). Hydrogen is relatively easily ignited by sparks.

Ignition Mechanism (4,7). Hydrogen, as we have already noted, ignites with a remarkably low ignition energy. It is also easily ignited with many other sources of energy, some of which are listed:

° Open Flames. Open flames are an obvious ignition source (others are subtle, obscure). One must control smoking, welding, burning and other flame sources in areas of potential hydrogen leakage.

° Electrical Equipment. Motors, lights, relays, switch gear can provide sparks of ignition intensity.

° Electrostatic Sparks. A spark of only 0.015 millijoules is sufficient to ignite hydrogen in air. Sparks of such intensity are easily generated in flowing hydrogen streams, either liquid or gas. The presence of a second phase in the flowing stream greatly enhances this charge separation process needed to produce these sparks. The second phase may be liquid droplets in a gas stream or solid particulate matter in a gaseous stream. The solid particulates may be gaseous impurities below their freezing points, such as nitrogen, oxygen, and carbon dioxide, or debris particles such as iron oxide or other dusts. Many hdyrogen fires have been initiated by electrostatic sparks.

° Sparks from Striking Objects. Sparks can be produced by metal-to-metal or metal-to-rock impact. Even spark resistant tools are not immune to spark generation.

° Thermite Sparks. The spark generation capability from striking of surfaces involving aluminum particles and iron oxide, because of the highly exothermic character of the $Fe_2O_3 + Al = Fe + Al_2O_3$ reaction, is substantial.

° Solid Air or Oxygen, Other Oxidants. Solid oxidants in liquid hydrogen can cause an explosion. The higher the enrichment of oxygen in the O_2-N_2 mixture, the greater the danger. Sparks will initiate a solid oxygen-liquid hydrogen explosion. A slurry of O_2 particles in liquid hydrogen will detonate upon initiation. The fracture of solid oxygen particles under liquid hydrogen is believed to provide sufficient energy under some conditions to initiate an explosion. Nitrous oxide, which can be present as a parts per million impurity in hydrogen obtained from certain electrolytic cells, will condense as a solid in hydrogen

liquefaction equipment below -132°F, and with very mild
ignition energy will detonate with high pressure gaseous
hydrogen down to -300°F.

° Hot Surfaces. Hydrogen-air mixtures will spontaneously
ignite when preheated into the 520-580°C range in the presence of stainless steel or other non-catalytic surfaces.

° Hot Hydrogen Leaking into Ambient Air. Hydrogen, above
about 680°C, will ignite if injected into ambient air. The
presence of dust or catalytic surfaces will lower this
temperature substantially.

° Catalytic Surfaces. Platinum and nickel catalysts will
bring about ignition of hydrogen-air mixtures at room temperature. Therefore, the possibility of surfaces having a
catalytic effect should be considered in evaluating ignition
hazards. Such surfaces could result in ignition of
hydrogen-air mixtures at any temperature from ambient to
the spontaneous ignition temperature.

Flash Arrestors. Flame arrestors prevent the passage of a
flame through the arrestor by removing sufficient heat from the
burning gas to lower the temperature below the temperature at
which a flame will propagate. Typical arrestors are of the
screen type and provide both rapid heat transfer through surface
area and heat sink through arrestor mass. In addition to the
screen type many other configurations will serve as long as
sufficient surface area and heat sink capability are provided.
Tube bundles, parallel plates and packed beds are other examples.
The quenching distance for hydrogen is much longer than for
hydrocarbon fuels because of its high flame speed and high
diffusivity making conventional screen type arrestors ineffective. Sintered bronze particles are effective with hydrogen.

Expansion of Liquid to Gas (8,9)

The critical temperature for liquid hydrogen is 33.19°K; it
cannot exist as a liquid above this temperature. If liquid
hydrogen is confined and then warmed to temperatures above the
critical point, very high pressures may occur far beyond the
working pressure of the hydrogen handling equipment. Figure 3
shows the density of para hydrogen as a function of temperature.
At atmospheric pressure the density of the vapor in equilibrium
with the liquid is 0.083 #/ft^3 (Point A), and the density of the
n.b.p. liquid is 4.43 #/ft^3 (Point B). If a line or vessel were
completely filled with n.b.p. liquid and that liquid trapped in
place the pressure will rise rapidly upon warming. This is read
from Figure 3 by tracing across to the right from Point B at the
constant density of 4.43 #/ft^3. At 120°R the pressure already
has risen to 5000 psi, at 200°R 10,000 psi and at 300°R 15,000
psi. At -160°F the pressure has risen to 15,000 psi. The pressure rise hazard for partially filled equipment is also signifi-

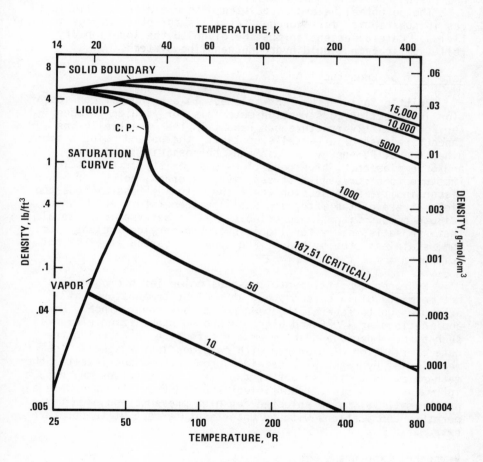

Figure 3.   Density of para hydrogen

cant. Assume that liquid hydrogen fills one-third of an apparatus at atmospheric pressure and that the apparatus is blocked
off. The average density will be about 1.5 #/ft³. At 140°R the
pressure has risen to 1000 psi, at 240°R 2000 psi and at 550°R
5000 psi. At a few degrees above room temperature the pressure
has risen from 14.7 to 5000 psi.

The potential hazard associated with expansion of liquid to
gas is substantial for hydrogen, as it is for other cryogenic
fluids. Careful attention must be given to the location of
relief devices in liquid hydrogen handling systems.

Materials of Construction

Brittle Failure (8). Brittleness is a principal consideration in selecting construction materials for liquid hydrogen
service. Brittle fracture can result in the essentially instan-
taneous release of a vessel's contents, the hazard being a combined one of PV energy release and the possibility of fire
and/or explosion. Three conditions must exist for a brittle
fracture to occur: 1) a stress riser, a crack, notch, or other
discontinuity, 2) a section where the actual stress exceeds the
yield stress of the material, and 3) a temperature below which
failure occurs without appreciable plastic deformation. Metals
that are satisfactory for liquid hydrogen service include aluminum, stainless steels, brass, and copper. Carbon steel is not
suitable.

Hydrogen Embrittlement. Gaseous molecular hydrogen does
not permeate plain carbon steel at ambient temperatures at
pressures up to several thousand pounds per square inch. For
applications at 400°F and higher, carbide stabilizing elements
such as molybdenum and chromium are effective in reducing the
rate of reaction of hydrogen with carbides in the steel. The
reaction of hydrogen with carbides produces methane in situ; the
methane can not diffuse out of the steel and stresses may result
in cracks or blisters. Sufficient knowledge exists to enable a
proper material to be selected for high pressure and high temperature applications with no uncertainty except at temperature
extremes.

Personnel Exposure

Cold "Burns" (8). Direct contact of body tissue with
liquid hydrogen for a short time may result in no damage because
the boiling liquid will be separated from the skin by a layer of
vapor and because hydrogen has a relatively low heat of vaporization. Cold gas jetting onto the skin can result in a high
heat flux sufficient to cause freezing. The body can tolerate
heat fluxes of 30 Btu/hr-ft² without discomfort; a flux of 740
Btu/hr-ft² will freeze facial tissue in about 10 seconds. A

more probable hazard is exposure to hydrogen handling equipment
cooled to some temperature intermediate between ambient and
n.b.p. hydrogen. Protective clothing is needed.

   Suffocation. A most unlikely hazard would be personnel
exposure to a hydrogen atmosphere sufficient to cause suffocation (a fire would be apt to occur first). However, it must be
borne in mind that hydrogen is as dangerous as the inert gases
in dilution of a breathing atmosphere.

   Diving Atmospheres (12). The U.S. Navy has conducted
research on pressurized breathing atmospheres for many years.
It is known that divers require roughly the same partial pressure
of oxygen at depth as is normal at the earth's surface, i.e.,
0.21 atmosphere. Somewhat higher partial pressures of oxygen
may be desirable at times (a football player breathing pure
oxygen for a few moments or a emphysema patient breathing in an
oxygen enriched environment) but the higher oxygen partial
pressures have the attendant hazard of causing higher burning
rates if a fire occurs. As the total pressure is increased,
while maintaining 0.21 atmosphere or some other relatively low
partial pressure of oxygen, nitrogen is no longer an acceptable
diluent for two reasons 1) nitrogen narcosis (rapture-of-the-
deep) and 2) the density becomes too great for normal lung work
behavior. Helium has been used to partially replace the nitrogen, trading the lighter gas with a molecular weight of 4 for
the heavier nitrogen with a molecular weight of 28. At depths
greater than 600 ft even helium-oxygen mixtures become too dense
for comfortable breathing. Hydrogen with only half the density
of helium has been suggested and tried by several divers at the
greater depths. The problems in using the H_2-O_2 mixtures are
simply those of avoiding an explosion while preparing the mixture
and avoiding a fire or explosion during use. It has been determined that a H_2-O_2 mixture will not burn if it contains less
than 5.3 volume percent oxygen. At depths of 200 ft and greater
the O_2 concentration can be lower than 5.3 volume percent and
still high enough to serve the divers needs. At even greater
depths the oxygen concentration will need to be much lower than
5.3 percent to avoid the toxic effects of oxygen (probably no
higher than 1.5 atmospheres partial pressure). Such H_2-O_2
mixtures will be inherently safe from the flammability viewpoint.

General Precautions and Safety Practices (2,7,8,10)

   Knowledge of the Hazards. Adequate knowledge of each of
the hazards associated with handling hydrogen in any form allows
all needed safety practices to be fully implemented. These
hazards have been identified as related to flammability, expansion from liquid to gas in confined spaces, improper materials
of construction, cold "burns," and breathing atmospheres con-

taining hydrogen. The reasons for concern for each of these
hazards have been identified and the technical character of the
hazards scoped. With such knowledge one may establish a system
for safe handling of hydrogen with high confidence that the
system may be operated without accident.

Assuming that a need for handling hydrogen exists, one
starts by gathering adequate knowledge on all the hazards of
concern. One then proceeds to design, construct, and operate
with hazard reviews at appropriate check points and with appropriate training and retraining of all involved personnel.

Safe Designs. Brittle fracture failure is a key consideration in materials selection. The system design will be planned
to avoid the entrance of oxidants into the system (or to achieve
their effective removal), to avoid leaks of hydrogen, and to
contain relief valves and rupture discs to protect all system
elements which could through misoperation become blocked-in
while containing liquid hydrogen. Grounding of equipment must
be practiced, remote shutoff valves may be installed, and special
attention will be given to system design to allow thorough
purging with particular attention to dead end lines (as to
pressure gauges) and pockets. This is important to avoid oxygen
contamination and also because essentially all impurities introduced in air are solid at liquid hydrogen temperatures and may
completely block passages in the system. Venting lines and
other lines which may carry high flow must be securely mounted
to avoid potential damage from a broken line whipping about
under the impulse of escaping gas. Detection instrumentation
for oxidants inside the system and for hydrogen outside the
system may be desired. If the system involves potential the for
oxidant accumulation (as solid oxygen) inside the system (as at
a filter/screen element), barricades for that system component
may be useful. Firefighting and other emergency equipment may
be planned into the system.

Safe Operating Practice. The system will have been designed
to allow safe operation. Operating Practices must be established
for normal operation, maintenance, and emergencies. The
Operating Practice should highlight all the practices related to
hazards, such as inspection of safety equipment, purging procedures, ventilation, barricade adequacy and use, exposure to cold
liquid, gas, or cold surfaces, and avoidance of flames and other
ignition sources.

Hazard Reviews. Formal and informal hazard reviews are
essential to establishing an operational history that is free of
accidents. Reviews should be made at key points during the
selection, design, construction, start-up and operation of a
system. Such reviews are cost-effective because they identify
hazards and permit problems to be solved before incorrect and

expense-incurring courses are followed. Frequency and formality are a function of the magnitude and complexity of the program which involves handling of hydrogen.

Training. Personnel should be fully informed of the character of the hazards for that portion of the system with which they are regularly involved and appropriately informed about the rest of the system. Training which results in a strong information base regarding the hazards, and a confident understanding of that information will allow good judgment to be exercised in emergencies.

Retraining and refresher courses are valuable.

Abstract

Safe handling of gaseous and liquid hydrogen is being accomplished through knowledge of its physical properties and its chemical reaction potential with oxidants, both solid and gaseous. The physical/chemical properties of hydrogen which must be understood to control hydrogen handling hazards are summarized. Hazards specifically discussed are gaseous hydrogen fires and explosions, liquid hydrogen explosions, cold "burns," embrittlement of handling systems resulting from low temperatures, bursting of equipment through evaporation of confined liquid hydrogen, and oxygen deficient breathing atmospheres. Criteria which may be used for hazard review of practices and procedures to control or avoid the listed hazards are provided.

Literature Cited

(1) Hord, J., "Is Hydrogen Safe." Cryogenics Division, Institute for Basic Standards, Nat'l. Bur. Stand., Boulder, Colorado, October 1976, NBS Tech. Note 690.

(2) McKinley, C., "Proceedings 1959 Safety Conference," sponsored by Air Products, Allentown, Pennsylvania, July 1959, 79.

(3) Coward, H. F., Jones, G. W., "Limits of Flammability of Gases and Vapors," Nat'l. Bur. of Stand., (1952), No. 503.

(4) Zabetakis, M. G., "Research on the Combustion and Explosion Hazards of Hydrogen - Water Vapor-Air Mixtures," OTS-U.S. Dept. of Commerce, AECU-3327.

(5) Zabetakis, M. G., "Flammability Characteristics of Combustible Gases and Vapors," Washington U.S. Dept. of the Interior, Bureau of Mines Bulletin 627, 1965.

(6) Lewis, B., vonElbe, G., "Combustion Flames and Explosions of Gases," 2nd Ed.; Academic Press., Inc.: New York, 1961.

(7) Weintraub, A. A., "Control of Liquid Hydrogen Hazards at Experimental Facilities," May 1965 Health and Safety Laboratory, New York Operations Office, U.S. Atomic Energy Commission.

(8) Vance, R. W., in "Applied Cryogenic Engineering," Ed.; Wiley & Sons, Inc., 1962; Chapter 10, C. McKinley.

(9) McCarty, R. D., "Hydrogen Technological Survey, Thermo-physical Properties," Cryogenics Division, Institute for Basic Standards, Nat'l. Bur. of Stand., Boulder, Colorado, 1975, NSAS SP-3089.

(10) Handbook for Hydrogen Handling Equipment, Arthur D. Little, Inc., 1960.

(11) Chemical Engineers Handbook, 4th Ed., McGraw-Hill, 1963.

(12) Dorr, V. A., Schreiner, H. R., "Region of Noncombustion, Flammability Limits of Hydrogen-Oxygen Mixtures, Full Scale Combustion and Extinguishing Tests and Screening of Flame-Resistant Materials," Ocean Systems, Inc., Tonawanda, N.Y., 1969.

RECEIVED July 12, 1979.

Production of Hydrogen for the Commercial Market: Current and Future Trends

C. R. BAKER

Union Carbide Corporation, Linde Division, P.O. Box 144, Tonawanda, NY 14150

The importance of hydrogen in the petroleum refining and chemical process industries is well established. Hydrogen is employed in a myriad of applications which range from the production of fertilizer to the upgrading of gasoline to the manufacture of semiconductors.

A list of some of the more important uses for hydrogen is given in Figure 1. The single greatest use is for the manufacture of ammonia, which consumes approximately 60% of all hydrogen produced; followed by hydrocracking of heavy residual oils to produce high quality gasoline, a process which consumes an additional 17% of total hydrogen production. Other important applications of hydrogen are for the hydrodesulfurization of sulfur-bearing petroleum streams and for the production of methanol, each contributing about 10%.

The total demand for hydrogen in the United States for the year 1975 has been placed (1) at a total of 0.58 x 10^{15} Btus (0.58 Quads), equivalent to a volumetric production rate of about 6 billion SCFD (Table I). The demand for hydrogen is expected to grow at an annual rate of about 4.5% to a level of 1.77 Quads by the year 2000. Others (2) have estimated the requirements for industrial hydrogen within the United States to be even greater, rising to as much as 3 Quads at the end of the century. These estimates do not include any new uses for hydrogen which may develop in the future, such as for the manufacture of synthetic fuels or for use directly as a fuel for automobile, rail, or air transportation. Such applications would be expected to increase the demand for hydrogen substantially, reaching levels as high as 14 Quads in the year 2000 and 50 Quads by 2020 (3). As a percentage of the total national energy demand, process hydrogen currently amounts to about 0.9, and should increase to 1.2 - 1.3% by the year 2000.

There are no significant quantities of naturally occurring free hydrogen anywhere on earth; hydrogen exists mostly in combined form. Major sources of hydrogen are shown in Table II. For commercial purposes, the two basic sources of hydrogen are

0-8412-0522-1/80/47-116-229$06.00
© 1980 American Chemical Society

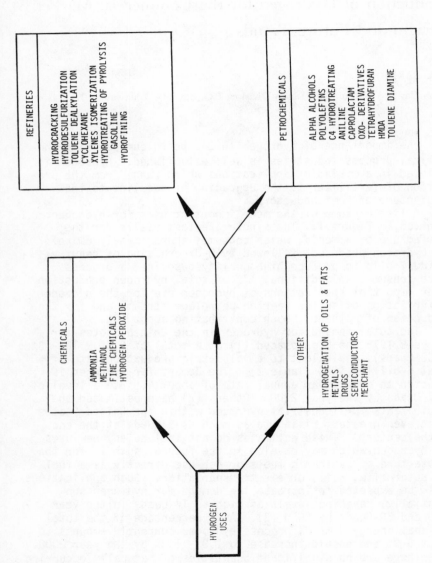

Figure 1. Uses for hydrogen

TABLE I.

FORECAST OF HYDROGEN REQUIREMENTS

QUADS*/YEAR

YEAR	1975	1985	2000
CHEMICALS			
AMMONIA	0.35	0.55	0.90
METHANOL	0.06	0.11	0.23
PETROLEUM REFINING	0.15	0.28	0.56
MISCELLANEOUS	0.02	0.04	0.08
TOTAL	0.58	0.98	1.77
U.S. ENERGY DEMAND	76.8**	98.5	140.7
H_2 AS % OF ENERGY DEMAND	0.76	0.99	1.26

* 1 QUAD = 10^{15} BTU = 3.64×10^{12} SCF

** FOR 1974

TABLE II.

HYDROGEN SOURCES

VIA GENERATION

STEAM REFORMING
- NATURAL GAS, LP-GAS, NAPHTHA

PARTIAL OXIDATION
- HYDROCARBON LIQUIDS, COAL

ELECTROLYTIC

VIA RECOVERY

REFINERIES
- CATALYTIC REFORMER OFF-GAS

PETROCHEMICALS
- ETHYLENE OFF-GAS
- STYRENE OFF-GAS

CHEMICALS
- AMMONIA LOOP PURGE
- METHANOL LOOP PURGE
- CHLORINE/CAUSTIC OFF-GAS
- SYN GAS - BY PRODUCT H_2 FROM CO RECOVERY
    PHOSGENE
    ACETIC ACID
    OTHER OXO-SYNTHESIS DERIVATIVES

METALS
- COKE OVEN GAS

water and hydrocarbons; and, of the hydrocarbons, methane is
predominant. Most of the commercially produced hydrogen is
derived from the catalytic reaction between steam and light
hydrocarbons, the well-known steam reforming process. In this
reaction, the hydrocarbon serves a dual role. It is not only
a source of hydrogen itself, but it is also an agent in the
thermal decomposition of water to remove and reject oxygen via
combination with the carbon atoms of the hydrocarbon molecule.
This is accomplished at temperature levels much lower than those
required for the unaided dissociation of water. In the reforming of methane, half of the hydrogen product is derived
from the steam. As the carbon to hydrogen ratio increases, so
does the proportion of hydrogen derived from the steam reactant.

Hydrogen Generation

     The steam reforming of natural gas is the economically-
preferred process in the United States for generating hydrogen
and synthesis gas today and, as a result, is responsible for
the bulk of the production of these gases. Commercial operation
of the catalytic steam reforming process began in 1930 but it
was in the period after World War II that, spurred by the
installation of extensive natural gas pipeline networks, major
growth and development took place. Up until the early 1960's,
only light hydrocarbon feedstocks such as natural gas and LP
gas were suitable for use in steam reforming. At that time,
a commercial process for the steam reforming of petroleum
naphtha was announced. Since then, naphtha has become a common
feedstock and several hundred such plants have been constructed
world wide.
     Partial oxidation processes for hydrogen generation are
capable of using, as feedstocks, hydrocarbon liquids which are
too heavy to be used in catalytic steam reforming. In this process, the endothermic heat of the reforming reaction is supplied
directly by combustion rather than by heat transfer across a
reactor tube wall. Burdened with the need for supply of oxygen,
partial oxidation plants have not been as widely used as steam
reformer plants and they contribute only a minor portion of the
total supply of hydrogen. As the supply of petroleum feedstocks dwindle and their price increases, partial oxidation of
hydrocarbon liquids is not likely to experience much future
growth and will remain a secondary source of hydrogen in the
years ahead. The partial oxidation process is advantageously
used in the production of low ratio ($H_2/CO = 1$) synthesis gas
required for manufacture of oxo alcohols and acetic acid.
     Coal gasification is also a partial oxidation process and
can be applied to production of hydrogen and synthesis gas. Although coal gasification processes have had considerable commercial success in Europe, Africa and Asia, they have only recently
begun to attract interest in the United States. The interest,

of course, has been spurred by the need to consider alternate
energy resources for replacement of oil and natural gas whose
limited proven reserves are rapidly being depleted. Resource
utilization based on coal is highly capital intensive and it
has been suggested (4) that for coal-based technology to be
widely adopted, investments for coal conversion techniques
and investments for finding new supplies of gas and oil must be
comparable. Investment for coal based production of hydrogen
is currently about double the investment for hydrogen derived
from natural gas and, therefore, feedstock conversion is not
likely to occur immediately and probably not until later in
this century. Earliest resource switching to coal is most
likely to occur in stationary fuel applications such as steam
raising and power generation because it can be accomplished
with the lowest investment, while the gas and oil resources
thus released will continue to be used for generation of hydrogen, and other chemical process needs.

Electrolysis of water is a minor source of hydrogen, providing less than 1% of the total requirements. It has to
support an investment burden even greater than coal based hydrogen and is, therefore, limited to applications where capital investment is a minor consideration, such as small capacity
plants, or where low cost electricity is available, such as
hydroelectric power. The bulk of the investment is related to
the generation of process energy and, before large capacity,
water-based electrolytic processes can become generally competitive, a reduction in energy investment is needed. An
increase in power generation efficiency would be of considerable
benefit in lowering the cost of electrolytic hydrogen.

Hydrogen Recovery

A large quantity of hydrogen is available via recovery
from by-product streams. Much of this hydrogen is unrecovered
and is simply burned as fuel. The largest source of recoverable
hydrogen is from catalytic reformer off gas streams in petroleum
refineries. These streams contain an amount of hydrogen equal
to about 75% of the nationwide total of newly generated hydrogen. Petrochemical plants for production of ethylene and styrene also produce hydrogen-rich off gas streams. In chemical
plants, purge streams from methanol and ammonia synthesis loops
contain hydrogen as does chlorine/caustic off gas streams.
Hydrogen is also available as a by-product in the recovery of
carbon monoxide from synthesis gas for manufacture of phosgene,
acetic acid and oxo-synthesis derivatives. In the metals industry, hydrogen accounts for approximately half of the total
composition of coke oven gas, from which it has been commercially recovered.

Hydrogen recovered from off gas streams characteristically
finds internal use at the source and major quantities of it are

recycled to process after appropriate treatment for upgrading
the hydrogen purity and rejecting the undesired components.
For example, total U.S. refinery requirements for hydrogen,
make up plus recovered, are about 4 billion SCFD of which
approximately half is consumed in residuum hydrocracking (1),
and the remainder mostly in desulfurization. Hydrogen obtained
from catalytic reforming and recycled provides about three-
fourths of the total refinery requirement while the remainder
is generated by steam reforming of natural gas and naphtha
plus a small amount by partial oxidation. United States
refinery hydrogen requirements are expected to grow at a rate of
about 5.5% annually for the remainder of the century.

Cryogenic Hydrogen Upgraders

Hydrogen contained in off gas streams is seldom, if ever,
of suitable composition and purity for direct recycle without
some kind of processing and it is appropriate to examine processing techniques in commercial use for the upgrading of
hydrogen from a variety of off gas streams. A very useful
instrument for accomplishing such objectives is the cryogenic
hydrogen upgrader. Table III presents a list of typical upgrader applications, each of which will be considered in some
detail.

TABLE III.

CRYOGENIC HYDROGEN UPGRADER APPLICATIONS

GENERAL HYDROGEN PURIFICATIONS
GRASS ROOTS OLEFIN PLANTS
AMMONIA VENT RECOVERY (AVR)
HYDRODEALKYLATION
SYNTHESIS GAS PROCESSING

There is an increasing incentive to recycle hydrogen
because of the upward spiraling costs of energy and hydrocarbon feedstocks. Low purity hydrogen is more valuable as a
chemical than it is as a fuel. It is more economic to cryogenically process hydrogen-containing off gas streams to recover a high purity (90-98%) hydrogen product than it is to
divert the off gas streams to fuel use and generate new hydrogen. The steam reforming of natural gas requires at least
0.45 cu.ft. of natural gas per cu.ft. of hydrogen produced,
equal to 425 Btu based on a 945 Btu/cu.ft. heating value for
the gas. With fuel valued at $3 per MM Btu, the fuel/feedstock
cost alone for producing the hydrogen is $1.28/MSCF. Deducting
an $0.83/MSCF credit for the hydrogen burned as fuel leaves

a net fuel/feedstock cost of $0.45 MMSC. In comparison, the total cost, including investment-related items, of cryogenically-upgraded hydrogen lies in the range of $0.05-$0.35 MSCF. With increasing costs of generating new hydrogen, the recovery of hydrogen from off gas streams for chemical applications will become more important. An additional benefit of the upgrading process is the recovery of light aliphatic and aromatic hydrocarbons at concentrations suitable for recycle to appropriate refinery and petrochemical operations. In many applications, the recovery of such hydrocarbons has completely justified the retrofit of an upgrader into an existing facility.

Description. The cryogenic hydrogen upgrader is represented schematically in Figure 2. In its basic concept, it consists of an assemblage of heat exchangers and phase separators arranged suitably within an insulated cold box to accomplish partial condensations and phase separations. The version shown in Figure 2 is a two-stage unit which cools the hydrogen-containing feed at a pressure of 300-625 psig from ambient temperature to a temperature level where a C_2^+ rich stream can be separated, and recycled to process or, alternatively, used as a medium pressure fuel. The vapor from the phase separation is further cooled to the lowest temperature level where a second phase separation produces a vapor fraction which consists of the upgraded hydrogen, and a liquid fraction which is primarily methane and which becomes a low pressure fuel stream. For purposes of refrigeration conservation, all separated streams are returned to ambient temperature in heat exchange with the feed stream.

Process variations can be incorporated into the upgrader to meet specific objectives. Figure 3 illustrates the use of a stripping column to remove light components from the C_2^+ fraction and produce a hydrocarbon stream more suited for recycle to process. Upgrader units normally operate in a self sufficient manner without need for an external source of refrigeration, and depend solely on the Joule-Thomson refrigeration inherent in the hydrocarbon components of the feed stream. Where this is inadequate, additional refrigeration can be provided by expanding a portion of the upgraded hydrogen through a cryogenic turboexpander.

Purities. Cryogenic hydrogen upgraders are capable of producing a hydrogen product at high recovery in the purity range of 90% to 98% at a pressure slightly below the feed pressure. Hydrogen purity is a function of feed pressure, temperature and composition at the final phase separation. Figure 4, presented for illustrative purposes, shows the hydrogen purity achievable for the separation of a 65/35 hydrogen methane feedstock with a final phase separation temperature established by the back pressure of the fuel (methane) stream.

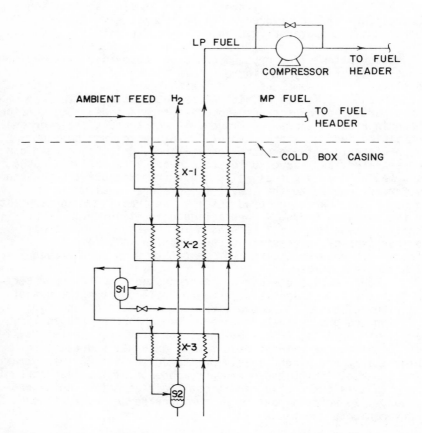

Figure 2.   Basic hydrogen recovery–upgrading process—the Joule Thomson cycle

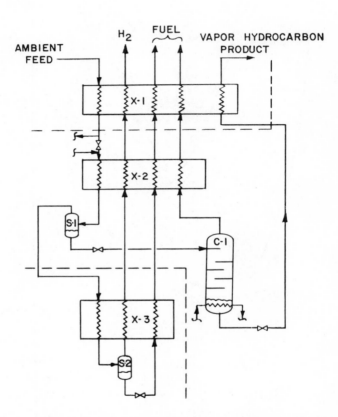

Figure 3.   Hydrogen–hydrocarbon recovery process with stripping column

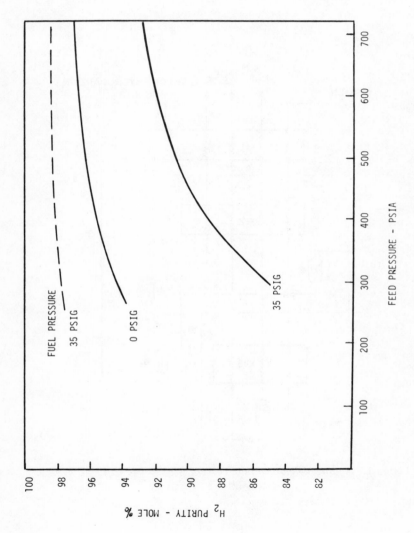

Figure 4. Cryogenic hydrogen upgrader performance—65/35 hydrogen–methane feedstock. Hydrogen recovery: (– – –), 95%; (———), 99%.

Hydrogen purity increases with increasing feed pressure, decreasing fuel pressure and decreasing hydrogen recovery. By sacrificing some hydrogen recovery, a small portion of the cold, upgraded hydrogen gas can be diverted into the low pressure liquid hydrocarbon stream for the purpose of decreasing its partial pressure and lowering the hydrocarbon boiling temperature. The resulting decreased temperature of the final phase separation produces a higher hydrogen product purity as illustrated by the broken curve in Figure 4.

Applications. An illustrative example of the application of upgrader technology is its integration into a hydrodealkylation unit. A principal commercial source of benzene is the dealkylation of aromatics such as toluene and pyrolysis gasoline fractions. A typical dealkylation schematic diagram is illustrated in Figure 5. A hydrogen/light hydrocarbons stream, after separation of the crude benzene product, is preprocessed for removal of water and aromatic content and then delivered to the cryogenic unit for upgrading. Processing accomplishes the separation of the feed stream into two major streams: a hydrogen product of 90-95% purity and a methane fuel stream. In addition, there may be minor product streams representing recovery of ethane and aromatics. The principal net accomplishment of the upgrader is the efficient rejection of methane from the dealkylation unit while minimizing overall hydrogen losses. The product hydrogen is mixed with makeup hydrogen and, after recompression, is recycled with toluene feedstock to the dealkylation unit. It is foreseen that the increased processing of pyrolysis gasoline feedstocks will represent most of the future growth in dealkylation processing.

In petroleum refineries, off-gas streams from catalytic reforming processes represent the largest source of recoverable hydrogen, exceeding by a wide margin the amount of make up hydrogen produced by steam reforming and partial oxidation. This source of hydrogen is being effectively utilized with the aid of the cryogenic hydrogen upgrader to recover and purify hydrogen for return to such refinery applications as residuum hydrocracking and hydrodesulfurization.

Olefin plants offer opportunities for cryogenic hydrogen upgrading systems either as a demethanizer feed chiller train or as a processing unit for demethanizer off gas streams. The latter application typically involves a retrofit installation at an existing plant for recovery of hydrogen and ethylene products while the former application would involve integration into a new plant at the design stage for the purpose of performing the several functions of a demethanizer feed chiller train listed in Table IV. In addition to performing the function of hydrogen recovery, the upgrader can be closely integrated into the olefins plant to provide for recovery of various hydrocarbon fractions and better utilization of plant

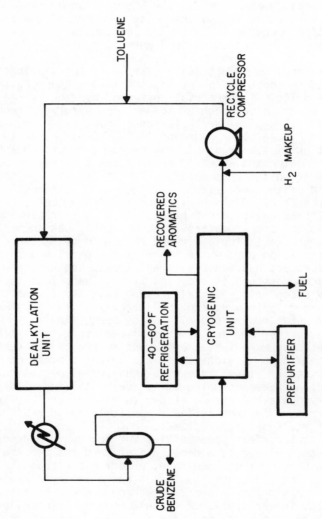

Figure 5.   Toluene dealkylation process

TABLE IV.

OLEFIN PLANT

APPLICATIONS

DEMETHANIZER FEED CHILLER TRAIN

1. RECOVERY OF ETHYLENE-RICH DEMETHANIZER FEED LIQUID STREAMS FROM CRACKED GAS FEED

2. REJECTION OF MULTIPLE FUEL STREAMS

3. RECOVERY AND UPGRADING OF HYDROGEN PRODUCT

4. REFRIGERATION RECOVERY

5. ETHYLENE REFRIGERANT CHARGE GAS CHILLERS

6. POTENTIAL UPGRADING OF LOW PURITY ETHYLENE STREAMS

DEMETHANIZER OFF GAS

1. RECOVERY AND UPGRADING OF HYDROGEN PRODUCT

2. RECOVERY OF ETHANE AND ETHYLENE

3. REJECTION OF METHANE FUEL STREAM

refrigeration. Demethanizer feed chiller trains represent one
of the larger and more significant applications of the cryogenic
hydrogen upgrader.

A further example of the application of hydrogen upgrading
technology is the processing of ammonia plant purge gas streams
via an Ammonia Vent Recovery (AVR) System. A schematic diagram
for the integration of an AVR System into a typical ammonia
plant is shown in Figure 6. The synthesis gas feedstock to an
ammonia plant typically contains, even after purification,
traces of argon and methane which are inert components in the
ammonia synthesis reaction. These tend to build in concentration in the synthesis loop and must be controlled by rejection
in a purge gas stream. The composition of this purge gas is
identical to that in the synthesis loop, i.e., mostly hydrogen
and nitrogen, and purging, therefore, normally represents a
diversion of a portion of the hydrogen to fuel. The use of a
cryogenic hydrogen upgrader to process the purge stream permits
the selective rejection of the inerts and recovery of the hydrogen for recycle to feed. Inclusion of a water scrubbing unit
before the cold box permits recovery of ammonia.

Pressure Swing Adsorption Systems

The Pressure Swing Adsorption (PSA) System is an alternative method to the cryogenic hydrogen upgrader for the separation, recovery and purification of hydrogen from a variety of
hydrogen containing process streams (5). The PSA process is an
ambient temperature, fixed bed, adsorption process, wherein the
adsorbed components are desorbed from the adsorbent bed by a
reduction in pressure rather than by an increase in temperature.
Because pressure reduction can be accomplished much more rapidly
than temperature increase, the length of the adsorption cycle
can be greatly decreased. The result is a reduced adsorbent
requirement and smaller adsorbent vessels with an accompanying
reduction in investment.

Description. The PSA cycle consists of four basic steps:
adsorption, depressurization, low pressure purge, and repres-
surization. The simplest system would consist of only two
adsorbent beds. One bed would be on stream in the adsorbtion
step while the other bed would be undergoing the remaining
three steps required for regeneration. The two bed system
suffers a severe disadvantage in the loss of a rather large
fraction of unrecoverable hydrogen during the depressurization
step. This disadvantage is overcome by use of a four bed
system, shown schematically in Figure 7. While bed 1 is on
stream, in the adsorption step, beds 2, 3, and 4 are in the
several stages of regeneration. Bed 2 is initially depressurized cocurrent to the hydrogen flow, into bed 4, which is being
repressurized. When pressure equalization is attained, bed 2

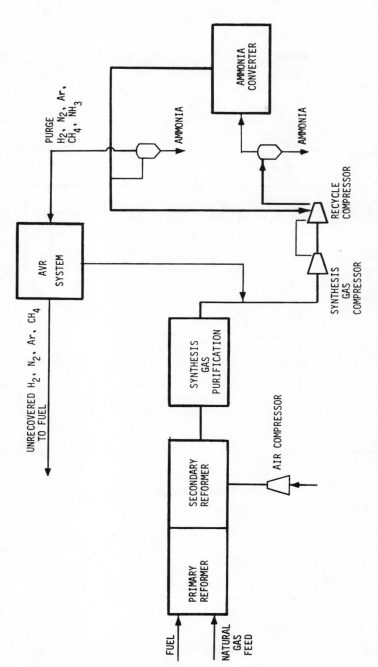

Figure 6.   Ammonia vent recovery system integrated with a typical ammonia plant

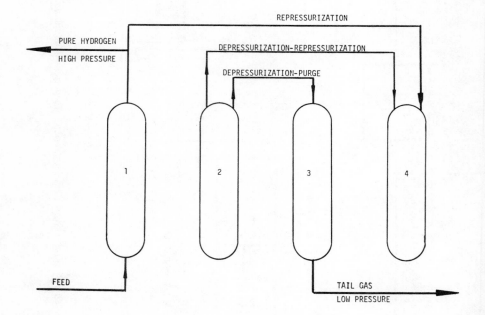

Figure 7.   Four-bed PSA system

is further depressurized cocurrently into bed 3 with the depressurization gas serving as purge. The desorbed components
are removed from the opposite end of bed 3, combined with the
purge gas, as a low pressure tail gas stream. Final depressurization of bed 1 occurs countercurrently. The cocurrent
depressurization step is the feature of the process which
allows more efficient use of the hydrogen stored in the ad-
sorber at the completion of the adsorption step and thereby
reduces hydrogen losses. Hydrogen depressurization losses are
decreased even further with the use of the Polybed PSA System
(6) which features more than four beds.

Although the PSA system of hydrogen recovery and purification is an alternative to the cryogenic hydrogen upgrader, it
has different characteristics and the systems are not completely
interchangeable. The PSA system produces hydrogen of extremely
high purity. Purity levels of 99.999% H_2 are normally achieved
and contaminant removal to levels of less than 1 part per
million has been commercially achieved for impurities such as
carbon monoxide, nitrogen and methane. This capability is not
inherent in the cryogenic hydrogen upgrader where purity levels
are normally 90% to 98% H_2. The upgrader, however, is a
higher recovery system than PSA. Although hydrogen recovery
factors are dependent upon process conditions and design
criteria, recovery from an upgrader is typically 95% or better
compared with approximately 80% in a 4 bed PSA system. In the
Polybed PSA, recoveries in the range of 85% - 95% are achievable.
Although capable of producing hydrogen of very high purity, the
PSA system is not limited to doing so. If desired, hydrogen of
lower purity can be produced while obtaining some improvement
in hydrogen recovery.

Applications. The range of applications for PSA system
parallels that for the cryogenic hydrogen upgrader. It has
particular applications where the exceptionally high purity of
the hydrogen product has added value. An application for the
use of a PSA system is presented in Figure 8, wherein hydrogen,
generated via steam reforming, is purified for use as a liquefaction feedstock. Impurity levels in the range of 1 ppm must
be maintained in order to prevent freezeup of the hydrogen
liquefier. The reformer effluent, after undergoing a single
stage of high temperature shift conversion is fed to the PSA
system. There, all the contaminating components of the feed
stream, including CO, CO_2, N_2, and CH_4 are removed and a pure
hydrogen product is delivered. The desorbed components, including some hydrogen, are mixed with make up fuel for recycle
to the burners of the steam reformer. Compared with a conventional purification train, the PSA system is a one-step purification process which effectively replaces the low temperature
shift converter, the CO_2 removal system and the methanator
while producing a much purer product than the conventional 97%

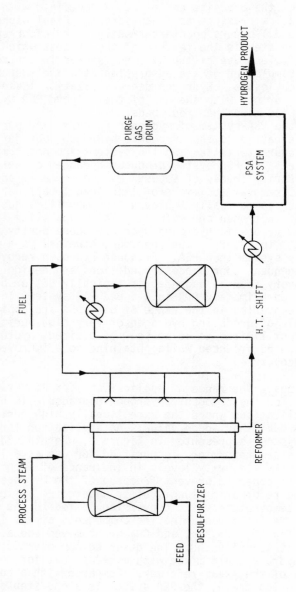

Figure 8. PSA hydrogen plant

level. In a comparative estimate (6), hydrogen was shown to
be purified more economically with a PSA system than with a
conventional purification train. In a similar application,
PSA would be used as a front-end purifier for the synthesis
of ammonia. The PSA system would produce, in a single purification step, an ammonia synthesis gas, free of the usual argon
and methane inerts; and, as a further advantage, would eliminate
the need for the purge stream from the ammonia synthesis loop.

Four-bed PSA units have greatest application in the 0.5
to 10 MMSCFD capacity range. Larger capacities justify the use
of the Polybed System. Fabrication of PSA units in single-train
capacities up to 40 MMSCFD of hydrogen product is presently a
commercial reality and capacities up to 60 MMSCFD are considered
feasible. They can, therefore, serve in large scale hydrogen
purification and recovery applications. For the future, they
will see increasing application in the purification of ammonia
synthesis gas and are likely to become a significant component
of coal gasification techniques used for production of chemicals
and synthetic fuels.

Combination Purification/Upgrading Systems

The combination of the cryogenic hydrogen upgrader and the
Pressure Swing Adsorption system can be advantageously applied
in certain situations to obtain higher hydrogen purities and
recoveries than is feasible when using either system alone.
The combination PSA/upgrader has special applicability for the
rejection of low boiling materials such as nitrogen, carbon
monoxide and/or methane when high recovery efficiency for
hydrogen is essential. An additional application is in the
processing of synthesis gas streams (7) to produce:
  ·   High purity carbon monoxide
  ·   High purity hydrogen
  ·   Ratio-adjusted synthesis gas
Carbon monoxide is an important building block in manufacture
of products such as phosgene - an intermediate in production
of isocyanates, polycarbonate resins, pesticides and herbicides;
and acetic acid - by direct reaction with methanol. Hydrogen
demand may be within or external to the chemical plant. An
example of internal use would be an integrated phosgene/iso-
cyanates facility where the hydrogen would be consumed in the
production of aniline or toluene diamine. The H_2/CO ratio of
the synthesis gas varies depending upon the oxo synthesis product being manufactured.

A schematic diagram for the processing of synthesis gas
to meet these three objectives using a combination upgrader/PSA
unit is presented in Figure 9. A portion of the synthesis gas
is diverted to the cold box for partial condensation and separation of the bulk of the methane and carbon monoxide from the
hydrogen. The enriched (94-98%) H_2 vapor fraction is reheated

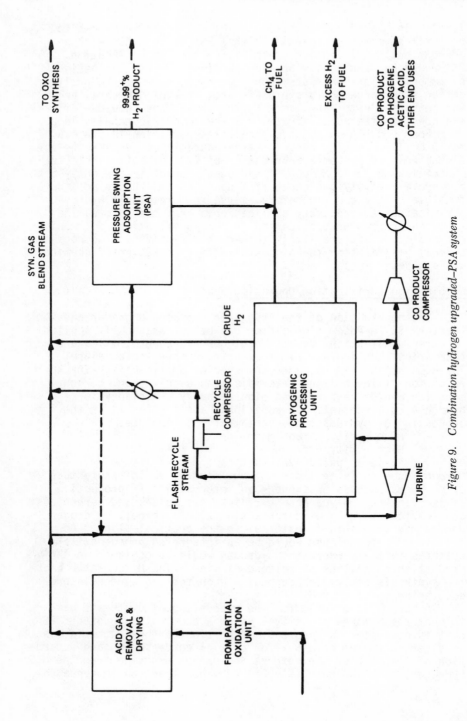

Figure 9.   Combination hydrogen upgraded–PSA system

and expanded to provide the refrigeration to sustain the process.
Unlike a hydrocarbon-rich off gas, a synthesis gas does not
possess sufficient Joule-Thomson refrigeration to overcome system
heat loads. The hydrogen is then reheated to ambient temperature
and part of it is used for blending and feed to the PSA unit.
Excess hydrogen may be used as fuel. The condensed liquid streams
are flashed to recover dissolved hydrogen from the CO-rich liquid
and then distilled to separate the carbon monoxide from the
methane. Both fractions are rewarmed in the cold box, the high
purity (98+%) carbon monoxide being recovered as product and
the methane being rejected for use as fuel. The flash hydrogen
from the cold box is compressed for blending into the synthesis
gas streams.

Hydrogen Use In Transportation

With world resources of petroleum feedstocks in limited
supply, future transportation needs for fuel are being jeopar-
dized. It is obvious that new sources of fuels must be provided.
As petroleum reserves continue to decline, hydrocarbons will
maintain their position as the preferred fuel for many applications although petroleum derived fuels will gradually be replaced
by alternates such as coal-derived synthetic liquids and possibly
LNG. A distinctly different alternate fuel which has recently
received considerable attention is hydrogen (8). Although
hydrogen has been proposed as a fuel for all transportation
sectors, it is foreseen that it will be applied initially in air
transportation. Conceptual design studies (9,10) made on hydrogen fueled commercial aircraft have shown considerable advantage,
not only for supersonic flight but also for subsonic aircraft in
comparison to their conventional Jet-A fueled counterparts.
The survey of the beginning of this paper predicted an 8-
fold increase in hydrogen demand at the turn of the century, and
a 12-fold increase twenty years later, if hydrogen should develop
as a fuel for transportation. This would have a major impact on
commercial hydrogen production in this country and is of sufficient significance to justify a more detailed examination in support of this forecast.
Hydrogen used as aircraft fuel must be stored on board in
liquid form because liquid is the minimum density state when the
storage container is included. Fuel production facilities would,
therefore, have to consist of not only the means to generate the
hydrogen but also the means for purification and liquefaction. A
typical purification-liquefaction complex (11) is shown schematically in Figure 10. The hydrogen would be generated using coal-
based technology, at least for the near future. NASA sponsored
studies (12,13) of airport facilities required for the support of
liquid-hydrogen fueled air transport placed average liquid hydrogen requirements at 650 TPD and 800 TPD, respectively, for two
large representative airports (San Francisco and O'Hare). The

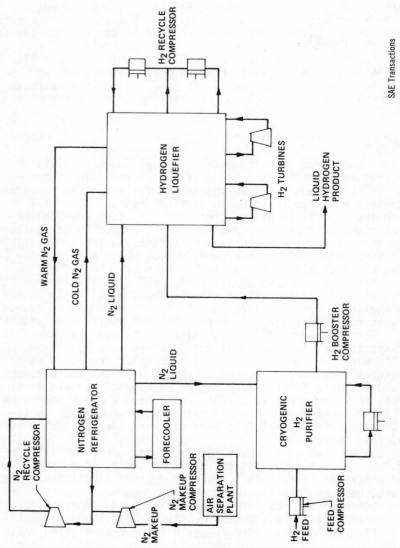

Figure 10. Block flow diagram for hydrogen purification–liquefaction complex (11)

SAE Transactions

coal requirements to produce the liquid hydrogen for these two airports would amount to 10 million tons annually of which 40% would be used to produce feedstock and the remaining 60% would be required to provide power. The coal consumed would amount to about 1.5% of our current annual production.

Total diversion of air transportation fuel to liquid hydrogen would represent a quantum jump in hydrogen demand. By 2000 A.D., the projected energy share for all transportation will amount to 30% of total energy consumption and the share for commercial aviation is expected to increase to 32% (14). Based on a total energy demand of 140 Quads in the year 2000, the amount of energy consumed by the aviation industry would be 13.4 Quads. This translates into 950 million tons of coal annually to provide the feedstock required for 135 MMSCFD (350,000 TPD) of liquid hydrogen plus an additional 1450 million TPY to provide the process energy. These requirements for liquid hydrogen are put into perspective by recognition that at the height of the Apollo Space Program, the total U.S. capacity for liquid hydrogen production was no more than approximately 200 TPD and that the typical plant capacity was 30 TPD. Clearly, the switch to liquid hydrogen as an aircraft fuel would require enormous commitment of both financial and energy resources.

There is every reason to believe that the transportation sector will become the major consumer of hydrogen in the next century and will require it in quantities which were undreamed of not too long ago. The need for processes and techniques to meet these demands must be recognized now so that necessary plans can be made to meet these demands efficiently and economically when the time arrives.

Abstract

The status of hydrogen in the U.S. market place is assessed in terms of hydrogen usage and applications as well as demand, both current and projected to the end of the century. Hydrogen is obtained by recovery from process off gas streams or by generation from water and hydrocarbons specifically for the intended end use. The relative importance of the several hydrogen producing processes is described as well as the future prospects for each. A more detailed examination is given to hydrogen recovery and purification technologies, specifically cryogenic hydrogen upgrading and pressure swing adsorption. Examples are given of the use of these technologies in actual present-day process applications. Finally, hydrogen as a fuel in transportation, especially air transportation, is considered and projected to result in major increases in hydrogen demand, requiring enormous commitment of both financial and energy resources.

Literature Cited

1. Corneil, H. G., Heinzelman, F. J., and Nicholson, E. W. S., Exxon Research and Engineering Co. - Brookhaven National Laboratories Report - 50663.

2. Kelley, James H. and Laumann, Eugene A., Jet Propulsion Laboratory, December, 1976, JPL 5040-1, Prepared under Contract No. NAS 7-100.

3. Rohrmann, C. A. and Greenborg, J., Int. J. of Hydrogen Energy, 1978, 2, 31.

4. Johnson, John E., Presented before the American Chemical Society National Meeting, March 13, 1978.

5. Wagner, J. L. and Stewart, H. A., "Paper Presented at the Novel Separation Systems Symposium, Third Joint Meeting, I.I.Q.P.R. and AIChE, May, 1970.

6. Heck, J. L. and Johansen, T., Hyd. Proc., 1978, 57, 1, 175.

7. Davis, J. S. and Martin, J. R., Presented at the 86th National Meeting, AIChE, April 1-5, 1979.

8. Gregory, D. P., Scientific American, 1973, 228, 1, 13.

9. Brewer, G. D., NASA CR114781, Prepared by Lockheed-California Company under Contract No. NAS2-7732, January, 1974.

10. Brewer, G. D., Morris, R. E., Lange, R. H., and Moore, J. W., NASA CR132559, Prepared by Lockheed-California Company and Lockheed-Georgia Company under Contract NAS 1-12972, January, 1975.

11. Baker, C. R., SAE Transactions, 1975, 83, 751094.

12. Brewer, G. D., NASA CR2700, Prepared by Lockheed-California Company under Contract NAS 1-14137, March, 1976.

13. Anon., NASA CR2699, Prepared by the Boeing Commercial Airplane Company under Contract NAS 1-14159, September, 1976.

14. Brewer, G. D., Int. J. Hydrogen Energy, 1976, 1, 1, 65.

RECEIVED September 17, 1979.

Hydrogen Distribution Safety

MANUS McHUGH III

Air Products & Chemicals, Inc., P.O. Box 538, Allentown, PA 18105

99% of all hydrogen consumed in the U. S. is produced on-site by users in the chemicals and refinery industries. Ammonia and methanol represent the greatest tonnage chemicals produced with on-site hydrogen. Other applications include metal processing, production of plastics and solvents, and hydrogenation of fats and oils.

Although representing less than 1% of current U. S. consumption, the liquid hydrogen industry is showing strong growth. This product is typically produced from a natural gas steam reforming facility, and it is cryogenically liquefied for ease of storage and delivery. In recent years, this maturing segment of the hydrogen industry is finding alternate applications to traditional rocket fuel use.

On-site storage and vaporization of liquid hydrogen is finding growth in such applications as: reducing atmosphere, reducing agents, high purity gas in electronic industry, hydrogenation, glass and gem manufacture, and on-site back-up during plant turn around.

There are five tonnage production facilities in the U. S. and one planned for Sarnia, Canada (Figure 1). Linde, a division of Union Carbide, has two plants, a 30 ton/day at Ontario, California, and a 17 ton/day at Ashtabula, Ohio. Airco has a 6 ton/day plant at Pedricktown, New Jersey. Air Products and Chemicals has two plants, a 60 ton/day at New Orleans, Louisiana (Figure 2) and a 30 ton/day at Long Beach, California; and also, a 15 ton/day planned for Sarnia, Canada.

The industry has developed a safe and reliable system for the distribution of liquid hydrogen. Although Linde trans-ships by approximately 30,000 gallon railcars and the government uses barges, the ultimate mode of transportation is by vacuum insulated tank trucks. The tank trucks range in size from 8 to 16 thousand gallons. Each complies with criteria established by the U. S. Department of Transportation. The industry has enjoyed an enviable safety record in the transport of liquid hydrogen. This is the result of attention paid to tanker design, testing, filling

0-8412-0522-1/80/47-116-253$05.75
© 1980 American Chemical Society

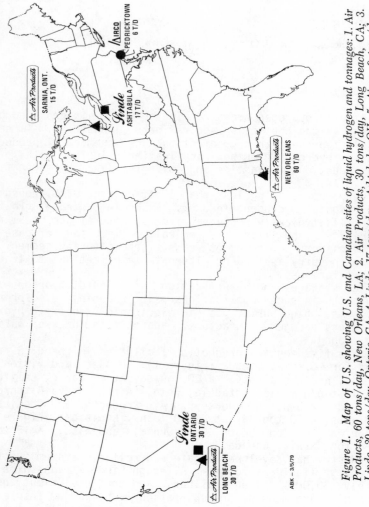

ABK – 3/5/79

Figure 1. Map of U.S. and Canadian sites of liquid hydrogen and tonnages: 1. Air Products, 60 tons/day, New Orleans, LA; 2. Air Products, 30 tons/day, Long Beach, CA; 3. Linde, 30 tons/day, Ontario, CA; 4. Linde, 17 tons/day, Ashtabula, OH; 5. Airco, 6 tons/day, Pedricktown, NJ; 6. Air Products, 15 tons/day, Sarnia, Canada.

Figure 2.   Air Products' New Orleans, LA, facility

at production facilities, and incorporation of safety systems to
meet the potential of vehicle malfunction and driver error. The
purpose of this paper is to discuss those safety considerations
built into the trailers. For ease of reference, Air Products
trailers and equipment will be used to illustrate my points,
(Figure 3).

The government regulations set forth the parameters for the
transportation of liquefied hydrogen which is classified as a
flammable gas and a hazardous material. These regulations are
embodied in special exemptions under which the industry is permitted to operate.

The corner stone of the exemptions is that the vehicles
shall be designed so as to permit no loss or venting of product
enroute. Since liquid hydrogen has a boiling point of approximately minus 423°F, the tankers are of a double-walled construction with the annulur space containing a multi-layered reflective
insulation operating under a high vacuum condition. Intrinsic to
the success of maintaining a high vacuum is the half inch material
used in construction of this outer shell and a well disciplined
program of vacuum maintenance. The support system is designed to
minimize heat input, and also to withstand load incident with
over-the-road transportation.

The design of these vehicles provides for a non-venting hold
time of up to 156 hours while building up a maximum pressure to
13 psig (Figure 4). To insure compliance with this requirement,
the driver monitors the vessel pressure every two hours. Additionally, a pressure (9 psig) warning system is visible to the
driver from his driving position (Figure 5). This system provides
advance warning of rising internal tank pressure (Figure 6) and
provides the driver adequate time to react. The maximum inner
vessel pressure may vary depending on the amount carried in the
trailer. In the full condition, the inner vessel pressure is
allowed to rise to 13 psig before the safety relief valve operates
or any venting is required. With a partial load (10,000 gallons
or less) or empty condition, the inner pressure is allowed to rise
to 50 psig before venting. The pressure relief valve settings and
related pressure/temperature/liquid expansion relationships establish the amount of pay load permitted. This is to prevent a
liquid full condition when the liquid is at vapor pressure equal
to the relief valve setting.

Each trailer is equipped with a dual safety valve and a dual
rupture disc system (Figure 7). These are independent systems
manually operated by the driver and can be switched to the alternate if the primary system fails. The capacities of the safety
valves are designed to handle a fire on the outside of the container with a uniform surface temperature of 1200°F.

The liquid hydrogen trailers are also designed and constructed to meet the rigors of highway operations and the conditions
encountered daily while transporting seven (7) billion cubic feet
of liquid hydrogen across the country per year.

Figure 3.   Air Products' tanker and equipment

Figure 4.   One-way hold time

Figure 5. A 9-psig warning system

*Figure 6. A 9-psig warning system as
seen through a tractor mirror*

Figure 7.   Dual safety valve and dual rupture disc system

The basics of any effective safety program for transportation is adequate training. Each driver is carefully selected and receives 18 hours of classroom training, 3 hours of video tape instruction, and an average of 60 days on the job training. The training is conducted by certified regional driver trainers. Management must personally qualify each driver before he is allowed to deliver liquid hydrogen. Systematic retraining and training reviews are carried out to insure that proficiency is maintained.

There are 76 individual steps in the pump delivery procedure (Figures 8 and 9) which requires the operations of as many as 40 valves. The importance of driver qualification is therefore obvious. In addition to normal operations, the driver must be able to perform under emergency conditions. Furthermore, drivers must be familiar with and observe all federal, state, and local regulations relative to the transportation of hazardous materials. They must also know the criteria outlined in the DOT exemptions issued for the transportation of liquid hydrogen.

The loading facility at Air Products New Orleans plant allows for the simultaneous loading of six trailers (Figure 10). Due to the light weight of liquid hydrogen (approximately ½ lb./gallon), the trailers are volume limited and not weight limited. Therefore great care must be taken to insure the trailers are properly loaded, allowing for sufficient vapor space, as outlined in the DOT exemptions. This vapor space allows for liquid expansion and proper hold time, preventing trailers from becoming liquid full as the product warms and expands. To prevent overfilling, the trailers are equipped with accurate liquid level gauges and a sensitive full trycock system. The full trycock senses the liquid level at a preset position (Figure 11), and, Air Products has developed a neon phase change device which is sensitive to the liquid hydrogen refrigeration. The device monitors the contents of the full trycock line and the gauge indicates when there is a presence of liquid hydrogen. Furthermore, it can automatically shut down the filling operations when the gauge indicates a presence of liquid hydrogen. Additionally, each trailer is carefully weighed before it is dispatched.

The fill (load) procedure employs a vapor return or vapor recovery system, which allows hydrogen vapor venting during loading (Figure 12), to be either returned to the hydrogen plant system or, if the plant is not operating, routed to remote vent system. This allows for an extra margin of safety while loading. In addition, the site is provided with purging and grounding equipment and a water deluge system (Figure 13).

Safety practices at the loading stations are maintained at very high levels. The vehicle ignition keys are removed from the vehicle; a system of warning flags, placed in the front of the trailer and at the driver's side rear (Figure 14), are employed to prevent accidental removal of a unit from the loading pad before filling operations are completed. All drivers are

LIQUID HYDROGEN PUMPING PROCEDURES

SHUT DOWN

1. Push stop button.
2. Close pump discharge valve.
3. Close EV 1.
4. Close V 42.
5. Close V 40.
6. Close V 41.
7. Open pump cool down
8. Open V 39.
9. Close bottom fill and trycock.
10. Open Hydrogen pump purge for 5 minutes.
11. Close top fill.
12. Open line drain.
13. Close pump cool down valve.
14. Open pump discharge.
15. Purge for 3 minutes.
16. Close Hydrogen purge valve.
17. Open Helium pump purge valve.
18. Set flow indicator to 90 CFH purge for
    10 minutes.
19. Vent trailer down to 3 psi if partial
    or empty.
20. Close Helium pump purge.
21. Close Helium seal shut off.
22. Close pump suction.
23. Close pump discharge.
24. Close valve on Helium cylinder.
25. Disconnect hose.
26. Close customer purge valve and replace
    dust cap.
27. Turn off electrical power and disconnect
    cord and put away.
28. Check trailer pressure and vent again
    to stabilize if necessary.
29. Close pressure building vapor valve.
30. Set pressure limiting control switch
    to full or partial accordingly.
31. Disconnect ground cable.
32. Pick up wheel chocks.
33. Close cabinet doors.
34. Make a complete circuit of vehicle
    before leaving.

Figure 8. Off-loading steps

LIQUID HYDROGEN PUMPING PROCEDURES

START UP

1. Check in with customer.
2. Spot trailer at customer tank, set brakes, put in low gear,
    take key out of ignition, chock wheels and ground trailer.
3. Put on gloves, goggles, and hard hat.
4. Turn full condition pressure switch to the "PARK" position.
5. Check customer tank for product integrity and discrepencies.
6. Clean connections, check O rings and teflon tips, hook up hose.
7. Open vent stack drain valve, then close.
8. Open customer purge line.
9. Check all valves on trailer, they should be closed, except V39
    close it at this time.
10. Open Helium cylinder.
11. Open pump suction and pump discharge.
12. Open helium pump purge valve.
13. Set Helium pump purge valve.
14. Purge for 3 minutes, for extended fill lines a longer purge
    may be required. Plug in electrical cord and turn power on.
15. Close customer purge valve.
16. Open pump cool down and purge trailer vent stack for 2
    minutes.
17. Close pump cool down.
18. Close Helium pump purge.
19. Open Helium seal shut off.
20. Set Helium flow indicator to 10 CHF.
21. Set EVI to normal operation.
22. Set V42 to open position.
23. Set V41 to open position.
24. Open vapor valve from pressure building coils.
25. Open liquid to pressure building coils and raise trailer
    pressure 10 psi or (20 to 25 psi).
26. Close liquid to pressure building coils.
27. Open Hydrogen pump purge valve.
28. Open customer purge valve purge for 3 minutes.
29. Close customer purge valve.
30. Close discharge valve.
31. Open customer top fill valve.
32. Open V40.
33. Crack open pump cool down valve until liquid air form on
    cool down line.
34. Open recycle valve.
35. Start pump. If it does not start in 10 seconds turn off
    and cool down again.
36. Close pump cool down.
37. Adjust recycle valve to maintain 140 psig to 150 psig.
38. Check trailer pressure maintain 10 psig above arrival pressure
    or (20 to 25 psig).
39. Crack discharge valve 5 seconds and close.
40. Open discharge and close recycle valve keeping pump pressure
    140 psi to 150 psi.
41. Open bottom fill if needed.
42. Open full trycock when customer tank is 75% full.

Figure 9. Off-loading steps

Figure 10.   View of loading facilities

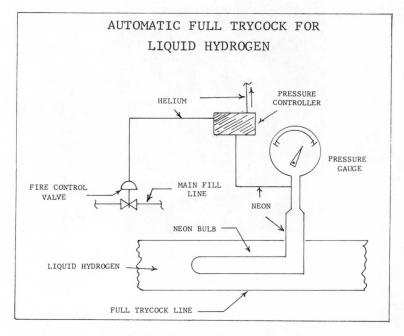

Figure 11.   Full trycock–neon bulb device

Figure 12.   Plant loading vent system

Figure 13.   Deluge system

Figure 14.   Warning flags

responsible for walking completely around the vehicle - to insure
that all valves are in their proper position, and all lines are
disconnected - before removing a unit from the loading station.
After the loading operation is completed, the vehicle keys are
returned to the driver and the warning flags are removed.

Predeparture Checks. Prior to and after loading, the drivers
must make a complete inspection of the vehicle to insure that it
meets all performance safety requirements of the federal government. A few of the numerous items that are checked are the lights,
tires, suspension, and brakes. Vapor pressure stabilization of
the liquid hydrogen is insured at this time, since venting of
hydrogen enroute is prohibited. Therefore, drivers insure that
the product is stabilized at its lowest possible pressure before
departing. To insure that the trailer's pressure is remaining
within prescribed limits, the drivers maintain a trip pressure
log every two hours during transit. At every two hour check, the
drivers also inspect the vehicle's tires, pressure, and condition,
and the trailer operating valves for tightness and position.

Unloading. Upon arrival at a delivery site, drivers check in
with the customer and then prepare to unload.
Static discharge grounding of the vehicle is one of the very
first steps taken in the loading/unloading operation. The liquid
hydrogen trailers are equipped with a fire control system (Figure
15) which halts the flow of hydrogen in the event of an emergency.
The fire control system consists of three fire control valves (one
on each liquid line), three emergency liquid shut off valves are
located in three places (Figure 16) on the trailer for safety and
ease of operation in an emergency. One is located in the operating side of the pumping control cabinet, a second is located on
the curb side rear trailer fender and the third is located on the
road side front trailer fender. Positioning any one of these
valves to the emergency shut off position automatically closes or
prevents operations of the three fire control valves.
The fire control valves are pressurized to open valves. The
drivers must activate the helium system which pressurizes the fire
control valves, opening them for delivery. The fire control
system can be pressurized when all emergency liquid shut off
valves relieve the helium pressure in the system, closing all the
fire control valves. Incorporated into the fire control system is
a pressure control monitor. If the inner vessel pressure reaches
a preset level during the unloading operation, the fire control
valve associated with the pressure building system automatically
closes. This prevents accidental overpressurization of the inner
vessel which could cause the relief of gaseous hydrogen during
delivery.
The liquid hydrogen trailers are equipped with a three way
full condition pressure limiting system (Figure 17) which coordin-
ates inner vessel pressure, in one of three modes, with the safety

Figure 15.   Fire control system

Figure 16.   Emergency shut-off switch

valve system and the vehicle brake system. When a full trailer is in transit, the 13 psig pressure limiting system is active, and the brakes are in the normal driver control operations. When a trailer is empty or partially loaded, the full condition pressure limiting system allows the 50 psig safety valve system to be activated, by-passing the 13 psig pressure limiting system, allowing for increased hold or travel time without venting. During delivery, the three way full condition pressure limiting system is switched to the park position, it locks the vehicle brakes and activates the 50 psig safety valve system. This permits the driver to manually raise the inner vessel pressure above 13 psig so that a positive inner vessel pressure above liquid saturation limits can be achieved for pump priming and off loading.

The actual unloading process is extremely involved and as pointed out above, only those drivers trained and qualified to deliver liquid hydrogen are permitted to operate these trailers.

Hydrogen gas or liquid is never allowed to mix with the atmosphere during the unloading operation (Figure 18). All trailer piping, the transfer hose, and the receiving customer station piping are purged with inert helium gas to insure all air is removed from the piping prior to introduction of hydrogen into the delivery system. The helium purge gas is then purged from the system with cold hydrogen gas from the trailer vapor space, then liquid hydrogen is introduced to the piping and pumping system.

During delivery, drivers constantly observe the pressure of the trailer inner vessel and the customer vessel. Great care is taken to insure that the customer station is filled at the pre-scribed pressure without overfilling. Here again, rigid training as well as a complete understanding of the properties of the product, and the use of safety equipment, such as safety hard hats, goggles, gloves, wheel chocks for the vehicle, and a grounding device to protect against static electricity are keys to proper unloading of liquefied hydrogen.

However, should there be an upset during unloading, the trailers are designed to safely vent hydrogen through the vent stack. As hydrogen can easily ignite, it is essential to be able to promptly and safely extinguish the fires, thus minimizing the interruption of the operations and exposure to fire damage. Air Products has developed a new fire extinguishing system (Figure 19) which constantly and dependably extinguishes vent stack fires without creating other hazards to personnel and equipment.

Customer Stations. Customer stations are product storage and supply systems which are usually permanently installed at consumer locations. These systems accept liquid product deliveries from the tankers and dispense gas at a controlled temperature and pressure to meet the customer's needs.

As the systems are unattended except during filling, safety is assured by installing the system in such a way as to minimize exposure and to comply with all applicable fire protection codes.

Figure 17.   Three-way valve

Figure 18.   Helium purge controls

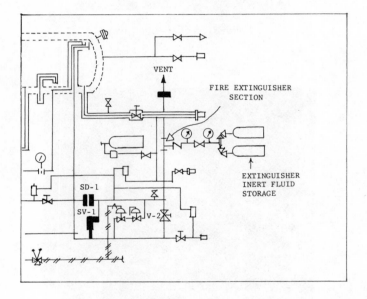

VENT

FIRE EXTINGUISHER
SECTION

EXTINGUISHER
INERT FLUID
STORAGE

SD-1

SV-1

V-2

Figure 19.   Drawing of vent fire control system

Figure 20.   Customer station

Figure 21.   Customer station controls

Figure 22.   Fire control valve to customer's houseline

Figure 23.   Air Products' tractor–trailer

A great deal of engineering review is given to each system before it is put on stream. A typical system (Figure 20) has a capacity from 1,000 to 20,000 gallons. The specific size is matched to the consumer demand to minimize vessel pressure build-up and possible venting of product to the atmosphere. Venting when it does occur is protected from any potential incident by an elevated vent stack which discharges at least 20 feet above grade. The stack is provided with a fire extinguishing system. Redundant safety valves and rupture discs (Figure 21) are supplied with selector valves to minimize product discharged from failure of a safety to reseat. Equipment to monitor vessel conditions is provided and installed in a manner that even if severe damage were to occur, the necessary information regarding the condition of the contents can still be gathered. Isolation of the system from the consumers houseline can be accomplished remotely with a fire control valve, (Figure 22), a mechanical failsafe device which can instantly interrupt the hydrogen liquid flame.

Detailed operating procedures have been established through careful study by engineering and operating personnel which promote an atmosphere of care and safety when the customer stations are handled beginning with start up. Initial start up and cooldown of each system is attended by an engineer and conducted in accordance with procedures which will minimize venting and assure that each component of the system is functional. In addition, scheduled preventative maintenance inspections are conducted by maintenance personnel to provide continuous assurance of equipment dependability.

Because of liquid hydrogen's extreme cold temperature, the equipment design must be of materials that have suitable properties for cold temperature operation. Vessels and piping are designed to ASME code and piping standards for the pressure and temperatures involved.

The hazards associated with liquid hydrogen are fire, explosion, and exposure to extreme cold temperatures. The elimination of sources of ignition minimize fire hazards. Careful purge operations performed by drivers using helium eliminate the formation of explosive mixtures during the start-up and shutdown of transfer operations.

Since 1952 Air Products has designed and built its own fleet of transport trailers for liquefied hydrogen (Figure 23). During the intervening years the company has operated over 30 million miles and delivered liquid hydrogen equivalent to more than 32 billion cubic feet of (standard pressure and temperature) hydrogen. In this time Air Products has not been involved in a serious accident where there was loss of product. The average mileage between accidents is 750,000 miles. This experience, combined with a constant awareness of safety, which is essential in any operation involving cryogenic fluids and flammable gases, enables Air Products to be the leader in supplying liquefied hydrogen safely and efficiently throughout the U. S.

RECEIVED July 12, 1979.

THE POTENTIAL OF FUTURE TECHNOLOGY AND APPLICATIONS

Hydrogen Requirements in Shale Oil and Synthetic Crude from Coal

J. L. SKINNER

ARCO Oil and Gas Company, Division of Atlantic Richfield Company,
P.O. Box 2819, Dallas, TX 75221

The development of a significant synthetic fuels industry
in the United States will probably occur first in the liquids
area. This is because the near-term outlook for increased
natural gas supplies is much more favorable than for increased
supplies of petroleum liquids. All indications are that a
world-wide, chronic shortage of petroleum liquids is imminent.
The delay in putting an alternate liquids industry on stream is
because the economic incentives, to date, have not been sufficient to warrant heavy front-end investments in ventures of
high technological risk. A very substantial cost element in
the production of shale oil and coal liquids is the cost of
hydrogen necessary to produce needed liquids, such as gasoline,
jet fuel, diesel oil, and clean fuel oils. This paper addresses
the question of the hydrogen requirements for producing useful
liquids from shale oil and coal.

Shale Oil

The hydrogen requirements for the upgrading of raw shale oil
are governed by one primary factor--its extraordinarily high
nitrogen content. An examination of Table I, which shows a
representative range of ultimate analyses for petroleum crudes
and for shale oils, shows that the carbon, hydrogen, sulfur,
and oxygen levels for raw shale oil are all within the range
typical for petroleum crudes. The nitrogen content of shale
oil, however, is twice that for high nitrogen crudes.
Shale oil is not a uniquely hydrogen deficient feedstock
vis-a-vis petroleum crudes. As can be seen from the representative data presented in Table II, its atomic hydrogen to carbon
ratio is within the range of some mid-continent crude oils. Its
atomic H/C ratio is that of a naphthenic base petroleum crude.
In order to produce acceptable liquid fuels from raw shale
oil most processing schemes incorporate either hydrotreating of

0-8412-0522-1/80/47-116-279$05.00
© 1980 American Chemical Society

TABLE I.

ULTIMATE ANALYSES OF CRUDE AND SHALE OILS

Molecule	Formula	% in Crude Oil	% in Shale Oil
Carbon	C	84-87	84.5-85.2
Hydrogen	H	11-14	11.2-11.3
Sulfur	S	0-3	0.64-0.76
Nitrogen	N	0-1	2.0-2.2
Oxygen	O	0-2	1.3-1.5

TABLE II.

ATOMIC H/C RATIO OF
REPRESENTATIVE CRUDE OILS AND SHALE OIL

Oil Type	Atomic H/C Ratio
Pennsylvania Crude	1.98
Healdton, Oklahoma Crude	1.81
Humbolt, Kansas Crude	1.73
Coalinga, California Crude	1.61
Beaumont, Texas Crude	1.53
SHALE OIL	1.58

the whole oil followed by subsequent processing, or fractionation of the whole oil followed by hydrotreating of the naphtha
and gas oil fractions. Regardless of the processing scheme,
hydrotreating for the removal of heteroatoms is required for
transportation fuels and may be required for fuel oils.

Although highly selective hydrodesulfurization can sometimes
be achieved, selective hydrodenitrogenation is much more difficult. The order of hydrotreating severity required for hydrogenation reactions is presented in Table III. As shown in this
table, if sufficient hydrotreating severity is employed to
remove most of the nitrogen, then hydrodesulfurization and
hydrodeoxygenation should be nearly complete. In addition, the
olefins will have been saturated. (Olefins are not normally
present in raw petroleum crudes, but they are present in raw
shale oil from a retort.) If sufficient hydrotreating severity
is employed to remove all, or nearly all, the nitrogen, then
substantial saturation of aromatics (including non-nitrogen
containing rings) will also occur. There will also be some
hydrocracking, although the hydrogen uptake for hydrocracking
will be substantially less than for aromatic saturation. Because
hydrodenitrogenation is not highly selective one can view Table
III as a series of overlapping bell curves with regard to hydrogen consumption. By the time one type of reaction is nearly
complete the next type of reaction is in substantial progress.

It is often useful to use a "model compound" approach when
attempting to analyze hydrogen consumption during hydrotreating
operations. Dineen, et al (1), report that the predominant
nitrogen compound types in shale oil are pyridines and pyrroles.
These heterocyclics are very stable and refractory to hydrogenation. Hence, severe hydrotreating conditions are required to
reduce nitrogen to acceptable levels.

Rollmann (2) has reported that with nitrogen and oxygen
species, saturation of any aromatic ring attached to the heteroatom is required prior to C-N or C-O bond scission. Thus, the
hydrogen requirement for denitrogenation of pyridines and
pyrroles is much higher than the hydrogen content of the ammonia
product from a hydrotreater. Figures 1 and 2 show the hydrogen
requirement for the hydrodenitrogenation of pyrrole and pyridine,
respectively. It can be seen that four hydrogen molecules are
required for the formation of one ammonia molecule from pyrrole,
and that five hydrogen molecules are required for the denitrogenation of a pyridine molecule. In an analogous manner one can
show that thiophenes will consume four molecules of hydrogen
for each molecule of hydrogen sulfide produced, and thioethers
will consume only two molecules of hydrogen per hydrogen sulfide
molecule. Rollmann has reported that saturation of aromatic
rings attached to sulfur species is not required prior to C-S
bond scission. However, at the high severities required for
hydrodenitrogenation, it seems reasonable to assume saturation
of the thiophenic rings.

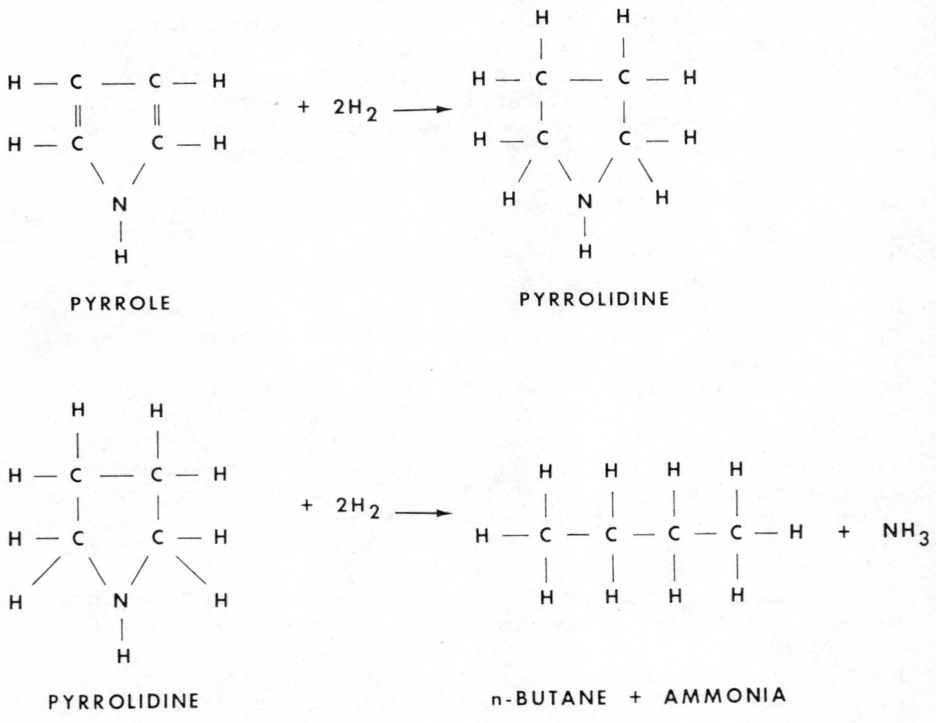

THUS, 4 H2 ARE REQUIRED FOR THE PRODUCTION OF ONE NH3

Figure 1.   Hydrodenitrogenation of pyrrole

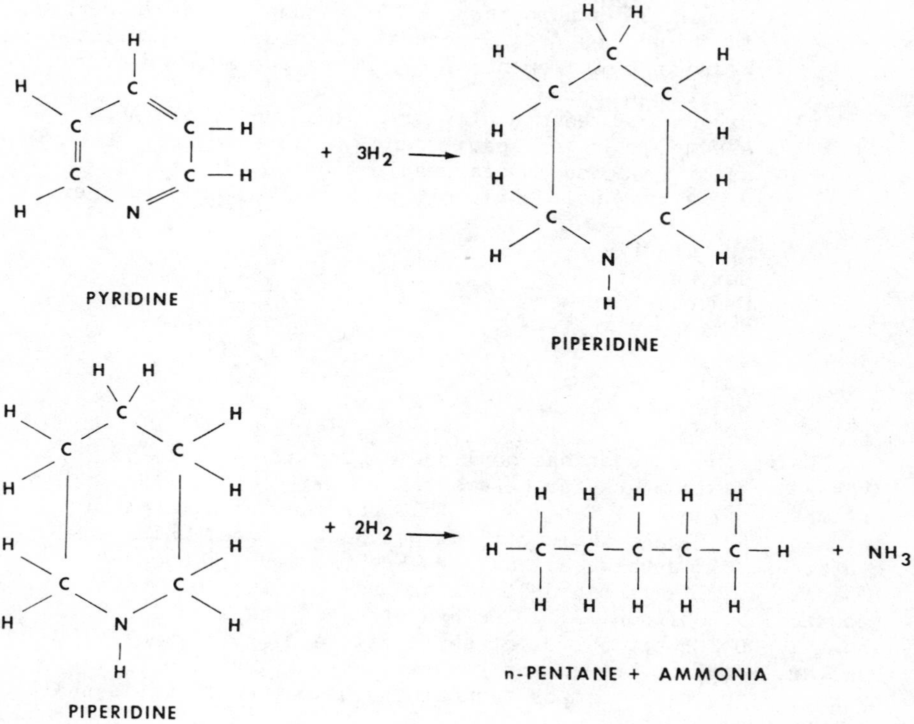

THUS, $5H_2$ ARE REQUIRED FOR THE PRODUCTION OF ONE NH_3.

Figure 2. Hydrodenitrogenation of pyridine

In order to make an a priori estimate of the hydrogen
requirement for heteroatom removal from raw shale oil the following assumptions were made:
  (1) Nitrogen is present in pyrroles and pyridines, with
       the nitrogen divided evenly between the two compound
       types. (This implies a hydrogen requirement of 4.5
       moles per mole of ammonia produced.)
  (2) Sulfur is present predominantly as thiophene, with the
       remainder as thioethers. The assumption of 75 percent
       in thiophenes and 25 percent as thioethers implies a
       hydrogen requirement of 3.5 moles per mole of hydrogen
       sulfide produced.
  (3) Oxygen is predominantly present as phenols. An
       assumed hydrogen requirement of 2.5 moles per mole of
       water produced is reasonable.
Consider a raw whole shale oil with the following properties:

API Gravity	20.3
Carbon, Wt %	84.71
Hydrogen, Wt %	11.32
Nitrogen, Wt %	2.14
Oxygen, Wt %	1.35
Sulfur, Wt %	0.68
Molecular Weight	297

Using the assumptions outlined above, an estimate of
hydrogen requirements for heteroatom removal can be made. This
estimate is shown in Table IV. This particular shale oil is,
in fact, the Paraho shale oil (direct heated mode) which was
hydrotreated for the U.S. Navy by The Standard Oil Company
(Ohio). Results from hydrotreating tests on this oil were
reported by Robinson (3). The hydrotreater was said to add
about 1,600 SCF per barrel of shale oil feed (12.05 kmol/m^3).
However, complete heteroatom removal was not achieved during
hydrotreating. The composition of the hydrotreated whole shale
oil was reported as:

Carbon, Wt %	85.93
Hydrogen, Wt %	12.96
Nitrogen, Wt %	0.30
Sulfur, Wt %	< 0.002
Oxygen, Wt %	0.53

Using the previously discussed assumptions it can be estimated
that another 220 SCF/bbl (1.66 kmol/m^3) of hydrogen would be
consumed in achieving complete heteroatom removal. Thus, the
results of the SOHIO study coupled with the above "model
compound" estimation process, indicates a hydrogen consumption
of about 1,820 SCF/bbl (13.70 kmol/m^3) for complete heteroatom
removal (neglecting the hydrogen uptake for non-hetero aromatic

TABLE III.

ORDER OF HYDROTREATING SEVERITY REQUIRED FOR
HYDROGENATION REACTIONS

Increasing
Severity

1. Saturation of Olefins

2. Desulfurization and Deoxygenation

3. Denitrogenation*

4. Saturation of Remaining Aromatics

5. Hydrocracking

*NOTE: Nitrogen containing aromatic compounds must be
saturated before the nitrogen can be removed.

TABLE IV.

ESTIMATED HYDROGEN REQUIREMENTS FOR HETEROATOM REMOVAL
FROM PARAHO SHALE OIL (DIRECT HEATED MODE)

	SCF/bbl	$kmol/m^3$
Denitrogenation	850	6.40
Desulfurization	90	0.68
Deoxygenation	260	1.96
Subtotal, Heteroatom Removal	1,200	9.04

saturation and hydrocracking associated with the removal of the remaining heteroatoms).

A similar approach can be followed when looking at the results of hydrotreating studies carried out by Chevron Research Company for the Department of Energy. These studies, which were carried out on a Paraho shale oil (indirect heated mode), were reported by Sullivan, et al (4). Using Chevron's results and "model compound" analysis one can estimate hydrogen consumptions of 1,874 SCF/bbl (14.11 kmol/m^3) and 1,925 SCF/bbl (14.49 kmol/m^3) for the two hydrotreating conditions reported. The somewhat higher consumption figure reported by Chevron may be due to the fact that they were processing a slightly different oil (average molecular weight of 326, as opposed to the average molecular weight of 297 of the oil processed by SOHIO).

What both the Chevron study and the SOHIO study indicate is that complete heteroatom removal from whole shale oil will require between 1,800 and 2,000 SCF of hydrogen per barrel of feedstock (13.55 kmol/m^3 to 15.06 kmol/m^3). What model compound analysis indicates is that hydrotreating of raw shale oil is rather selective, in that only about one-third of the hydrogen consumed is for olefin saturation, saturation of non-hetero aromatics, and hydrocracking.

A simplified processing scheme for fractionation of the whole oil followed by hydrodenitrogenation of the naphtha and gas oil cuts is shown in Figure 3. This approach minimizes potential problems with hydrotreating at the expense of overall liquid yield. There are two sources of hydrogen in this scheme-- the retort gas stream and the bottoms. The bottoms can be fed to a coker in order to make more liquids and a coke product, or they can be fed to a partial oxidation unit to produce syngas. The final selection of a processing scheme will involve not only bringing the shale oil plant into energy and hydrogen balance, but also will involve economic optimization with regard to coke and pipeline gas.

A simplified process diagram for hydrotreating of whole shale oil followed by fractionation and subsequent processing is shown in Figure 4. This approach maximizes the possible yield of usable liquids at the expense of "overhydrotreating" of some of the light and middle distillate cuts. What is common, however, with the scheme discussed previously is that the potential hydrogen sources are in the gas and the bottoms. Those liquids streams in the middle are far too valuable to be considered as hydrogen sources. Again, market considerations and economic optimization will determine the method of hydrogen production and the split of gas and bottoms between hydrogen production and plant fuel. In all cases, the market factors will be influenced by geographic locations, and environmental considerations will impact on the selection of processing schemes.

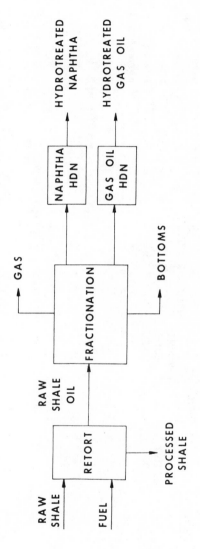

Figure 3.   Fractionation followed by hydrotreating

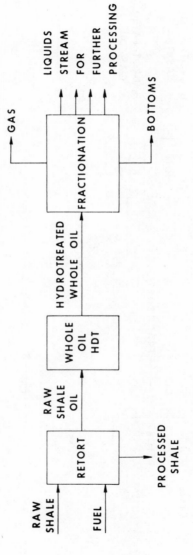

Figure 4.   Hydrotreating followed by fractionation

If shale oil is desired as a fuel oil rather than as a feedstock for transportation fuels, it may be possible to avoid hydrotreating (and hydrogen production) altogether. Southern California Edison has reported that raw shale oil could be burned in a "Dual Fuel Combustion" boiler without exceeding nitrogen oxide emission limits (5). Shale oil was fed to burners in the lower section of the boiler at off stoichiometric conditions. Gas was burned in the upper section. Up to 58 percent of the energy was supplied as shale oil without exceeding the NOx emission limits. Blending of shale oil and petroleum derived low sulfur fuel oil was a less effective method of utilizing raw shale oil. Nitrogen oxide emissions could only be met with blends of 17 percent shale oil, or less.

Coal Liquids

Coal, as opposed to raw shale oil, is hydrogen deficient relative to petroleum crudes. Representative comparisons of coal raw shale oil and petroleum crudes are shown in Table V. An examination of this table shows, however, that not only are coals hydrogen deficient, they also contain higher concentrations of heteroatoms (N, S and O) than shale oil. Therefore, heteroatom removal comprises a large fraction of the hydrogen consumption, although not to the same extent as shale oil.

Actual measurements of the hydrogen uptake necessary for coal liquefaction show that between 5,000 and 6,500 SCF per barrel (37.64 to 48.94 $kmol/m^3$) are required to produce fuel oils and that between 7,000 and 8,000 SCF (52.70 to 60.23 $kmol/m^3$) are required to produce material of a synthetic crude quality. The conversion of distillate cuts of this syncrude to usable transportation fuels requires another 1,000 to 3,500 SCF/bbl (7.53 to 26.35 $kmol/m^3$), depending upon the quality of the coal liquid feedstock (6). A reasonable estimate of the total hydrogen requirement for the production of a barrel of coal-derived transportation fuel is about 10,000 SCF (75.29 $kmol/m^3$). The larger the hydrogen uptake during liquefaction, the smaller the hydrogen requirement for subsequent upgrading. There is little likelihood that the entire boiling range of coal liquids produced by direct liquefaction (SRC-II, H-Coal, Exxon Donor Solvent, etc.) will be used as transportation fuels feedstock. The upgrading costs associated with the higher boiling cuts would be enormous. These cuts will be used for fuel oils or hydrogen production, with extensive upgrading reserved for the lower boiling range materials.

The source of hydrogen for coal liquids production could be, as in the case of shale oil production, either the gas or the bottoms product from the liquids plant. Again, the choice of feedstock for hydrogen production will be dictated by economic, market, and environmental considerations.

TABLE V.

COMPARISONS OF PETROLEUM CRUDE, SHALE OIL, AND COAL

	Atomic H/C Ratio	Heteroatoms Content (N,S,O) Wt %
Petroleum Crudes	1.5-2.0	1-6
Raw Shale Oil	1.58	4-4.5
Lignite	0.98	16-20
Typical Bituminous Coal	0.84	12-16
Low Volatile Bituminous	0.61	6-12
Anthracite	0.34	4-6

Abstract

Total hydrogen requirements for liquid fuels derived from
oil shale and coal are dependent upon the feed raw material and
upon the desired liquid product. For the production of a
synthetic crude product from oil shale, about 2,000 SCF of
hydrogen per barrel (15.06 kmol/m^3 is required; whereas the
production of synthetic crude from coal will require between
7,000 and 8,000 SCF per barrel (52.70 to 60.23 kmol/m^3). The
hydrogen requirements for a fuel oil product are less than for
a syncrude product. For fuel oil produced from coal, the
hydrogen requirements will range between 5,000 and 6,500 SCF
per barrel (37.64 to 48.94 kmol/m^3), depending upon the liquefaction scheme. In the case of shale-derived fuel oil it may be
possible to blend off the raw shale oil (or lightly hydrotreated
shale liquids) with petroleum derived fuel oil and thus minimize
"on-site" hydrogen requirements.

In all cases, total plant hydrogen demand will dictate a
dedicated hydrogen plant which is designed and built as an
integral part of the syncrude or fuel oil plant. In all probil-
ity, no one process for hydrogen production will be dominant
in first generation commercial plants. By-product prices,
geographical location, and environmental considerations can all
influence which "by-product stream" in the plant is used for
feedstock to the hydrogen plant.

Literature Cited

1. Dineen, G. U., Cook, G. L., Jensen, H. B., Anal. Chem. 30,
    2026 (1958).

2. Rollmann, L. D., Journal of Catalysis 46, 243-252 (1977).

3. Robinson, E. T., "Composition of Shale Oil and Shale Oil
    Derived Fuel," Conference on the Composition of Transporta-
    tion Synfuels, Southwest Research Institute.

4. Sullivan, R. F., Strangeland, B. E., "Converting Green River
    Shale Oil to Transportation Fuels," 11th Oil Shale Symposium,
    Colorado School of Mines, (1978).

5. Mansour, M. N., Jones, D. G., "Emission Characteristics of
    Paraho Shale Oil as Tested in a Utility Boiler," Electric
    Power Research Institute Report AF-709, March 1978.

6. Heck, R. H., Stein, T. R., "Kinetics of Hydroprocessing
    Distillate Coal Liquids," Symposium on Refining of Synthetic
    Crudes, Chicago, (1977).

RECEIVED September 24, 1979.

15

Rechargeable Metal Hydrides: A New Concept in Hydrogen Storage, Processing, and Handling

G. D. SANDROCK

The International Nickel Company, Inc., Inco Research & Development Center, Sterling Forest, Suffern, NY 10901

E. SNAPE

ERGENICS Division, MPD Technology Corporation, 4 William Demarest Place, Waldwick, NJ 07463

Hydrogen has been traditionally stored, transported, and used in the form of compressed gas or cryogenic liquid. The purpose of this paper is to discuss a third alternative, namely the use of rechargeable metal hydrides.

In the most elementary sense, a rechargeable metal hydride is a metal powder that can act as a solid "sponge" for hydrogen. In a chemical and practical sense, of course, the concept is a bit more complicated than that. In this paper we will present an introduction into the science and applications of rechargeable metal hydrides. It will consist of three parts: (a) a review of the fundamentals and practical properties of metal hydrides, (b) a survey of the main families of rechargeable hydrides that are commercially available and, (c) a brief summary of the potential applications of hydrides within the existing hydrogen industry.

Fundamentals of Rechargeable Metal Hydrides

Basic Chemistry and Thermodynamics. The key to the understanding and use of rechargeable metal hydrides is the simple reversible reaction of a solid metal Me with gaseous H_2 to form a solid metal hydride MeH_x:

$$Me + \frac{x}{2} H_2 \rightleftarrows MeH_x \qquad\qquad (Eq. 1)$$

Not all metals react directly with gaseous H_2, and of those that do, some are not readily reversible (i.e., the reverse reaction to liberate gaseous H_2 cannot be readily performed). Fortunately, there are a number of metals that do react directly and reversibly in the manner of Eq. 1, and do so at practical temperatures and pressures (e.g., room temperature and near atmospheric pressure). Such metals include elements, solid-solution alloys, and especially intermetallic compounds. We call the hydrides of these metals "rechargeable metal hydrides". In effect, the metal becomes a solid "sponge" for hydrogen that can be repeatedly charged and discharged at will. They bear a physical analogy

0-8412-0522-1/80/47-116-293$07.50
© 1980 American Chemical Society

with a water sponge and chemical analogy with a rechargeable
electric battery.

Rechargeable metal hydrides offer a number of advantages
over compressed gas and cryogenic liquid for the storage and
handling of hydrogen in the commercial sector. A primary
advantage of hydrides is their extremely high volumetric packing
density for hydrogen(1). As shown in Table I, the volumetric
density of H in typical hydrides is many times that of high
pressure gas and even significantly greater than that of liquid
hydrogen. This is a result of the fact that H-atoms are chem-
ically bound very compactly within the hydride crystal lattice.
A second advantage of hydride storage is the low pressures that
are required, a factor that has safety implications to be discussed in more detail later. A third advantage, especially
relative to liquid hydrogen, is the high energy efficiency of
hydride storage. This, again, will be discussed in more detail
later in this paper. Disadvantages of hydrides, such as cost
and weight, will also be discussed in the appropriate context.

Table I. Hydrogen Content of Various Media (1).
(Does not include container weight or void volumes)

Medium	Wt.% H	Volumetric Density N_H (Atoms H/ml, x 10^{-22})
H_2, liquid	100	4.2
H_2, gas at 100 atm	100	0.5
MgH_2	7.6	6.7
UH_3	1.3	8.3
TiH_2	4.0	9.1
VH_2	2.1	11.4
$Mg_2NiH_{4.2}$	3.8	5.9
$FeTiH_{1.74} \rightarrow FeTiH_{0.14}$	1.5	5.5
$LaNi_5H_{6.7}$	1.5	7.6

The phenomenological and thermodynamic aspects of the
reaction shown in Eq. 1 should be discussed in more detail.
Typically, the absorption and desorption properties of metals
are determined from pressure composition (P-C) isotherms, an
idealized form of which is shown in Figure 1. If we start with
the metal phase at Point 1, maintain a constant temperature and
slowly increase the H_2 pressure, relatively little happens at
first. As the H_2 pressure increases, a small amount of hydrogen
goes into solution into the metal phase. At some pressure
(Point 2), the hydriding reaction (Eq. 1) begins and the sample
starts to absorb large quantities of hydrogen at nearly constant
pressure. This pressure, P_p, is called the "plateau pressure".
The plateau 2-3 then corresponds to a two-phase mixture of metal
Me and metal hydride MeH_x. At Point 3, the sample has been
completely converted to the hydride phase and a further increase

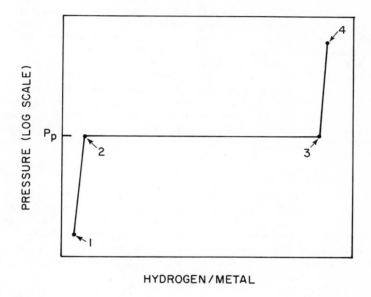

Figure 1. *Ideal absorption and desorption isotherm for a metal–hydrogen system*

in applied H_2 pressure to Point 4 results only in a small additional pickup of hydrogen in solution in the hydride phase. In principle this curve is reversible. As H_2 is extracted from the gas phase in contact with the sample, the hydride phase will dissociate (dehydride) to Me + H_2 gas and attempt to maintain the equilibrium plateau pressure until it is fully dissociated (back to Point 2).

Although the ideal behavior shown in Figure 1 is occasionally observed in practice, there are usually slight deviations from this ideality. To illustrate this, isotherms obtained for a practical nickel-aluminum-mischmetal compound (2) are shown in Figure 2. Compared to the ideal curve (Figure 1), the plateau is often sloped slightly and the plateau limits are often not as sharp. In addition, there is almost always some pressure hysteresis between absorption and desorption (see the 25°C curves of Figure 2). The hysteresis in this example is relatively small, although clearly measurable.

Figure 2 shows the strong dependence of temperature on the plateau pressure. The higher the temperature, the higher is the plateau pressure. This is an important thermodynamic consequence of the heat of reaction ΔH associated with Eq. 1. The hydriding reaction($\rightarrow$) is exothermic and the dehydriding reaction($\leftarrow$) is endothermic. The plateau pressure P_p is related to the absolute temperature T by the Van't Hoff equation

$$\ln P_p = \frac{2}{x} \frac{\Delta H}{RT} + C \qquad \text{(Eq. 2)}$$

where x is defined in Eq. 1, ΔH is the enthalpy change (heat) of the hydriding reaction, R is the universal gas constant, and C is a constant related to the entropy change of the hydriding reaction. Thus, from a series of experimental isotherms such as shown Figure 2, a Van't Hoff plot of ln P_p vs. 1/T can be made and the value of ΔH for a particular material can be readily determined from the slope of that plot.

This is done in Figure 3, using the H/M = 0.5 values of P_p taken from the desorption isotherms in Figure 2. The resultant value of ΔH is -6.7 kcal/mol H_2 for this particular alloy. This is the heat that is generated during the hydriding reaction and must be supplied during the dehydriding reaction. Note that the heat involved in this case is only about 12% of the lower heating value of the hydrogen involved (-57.8 kcal/mol) and represents a "low-grade" form of heat. An inspection of Figure 2 or Figure 3 shows that room temperature (or even ice water temperature) "waste" heat would be capable of dissociating $MNi_{4.5}Al_{0.5}$ hydride to provide H_2 at pressures of 1 atm absolute or more.

Engineering Properties. In addition to the simple principles discussed above, there are a number of engineering properties that bear on practical applications of hydrides in the hydrogen storage, processing, and handling fields. Most of these are

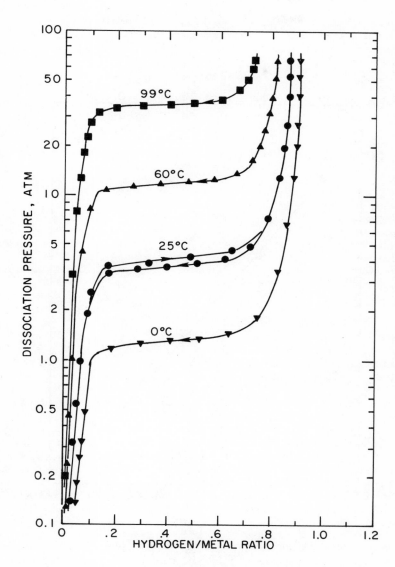

Figure 2.   Various isotherms for MNi$_{4.5}$Al$_{0.5}$ annealed 4 hr at 1125°C (2)

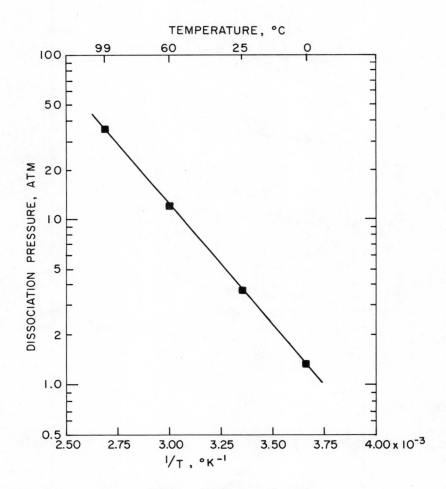

Figure 3. Van't Hoff plot (desorption) for annealed $MNi_{4.5}Al_{0.5}$. $\Delta H = -6.7$ kcal/mol H_2 (2).

listed in Table II and will now be discussed briefly.

Table II.   Important Engineering Properties of
Rechargeable Metal Hydrides

Plateau Pressure/Temperature
Plateau Slope
Hysteresis
Heat of Reaction
Hydrogen Capacity
Volume Change
Rate of Decrepitation
Ease of Activation
Kinetics of Reaction
Tolerance to Gaseous Impurities
Chemical Stability (Disproportionation)
Thermal Conductivity
Specific Heat
Safety
Production Factors
Cost and Long Term Availability of Raw Materials

Plateau pressures or temperatures desired depend on the
application intended. The plateau pressure at a given temperature is a strong function of the metal composition. The state of
the art has expanded greatly in the last few years, so that we
now have a wide variety of hydrides available and can, in many
cases, tailor-design plateau pressures or dissociation temperatures. We will survey the main classes of hydrides later. As a
preview to that survey, desorption Van't Hoff curves of just a
few representative materials are presented in Figure 4 to show
the wide range of materials that are available in the 300°C to
-20°C range.

Plateau slope is usually a function of metallurgical segre-
gation that sometimes occurs during the production of a hydriding
alloy(2,3,4). In some applications a sloped plateau is desired,
but in most cases a reasonably flat plateau is preferred. We
have learned to vary (or largely eliminate, if desired) the
plateau slope by heat treatment techniques(2,4).

Hysteresis, the pressure difference between absorption and
desorption, should be small for most practical applications of
hydrides. Hysteresis varies markedly from alloy to alloy in a
manner not understood by researchers in the field. However, an
empirical body of knowledge now exists to show a number of
systems where hysteresis can be made quite small. The nickel-
aluminum-mischmetal system (e.g., Figure 2) is a good example of
one with small hysteresis.

Heat of reaction is one of the most important hydride properties from a container design point of view. Because heat is
generated during hydriding and required during dehydriding, the

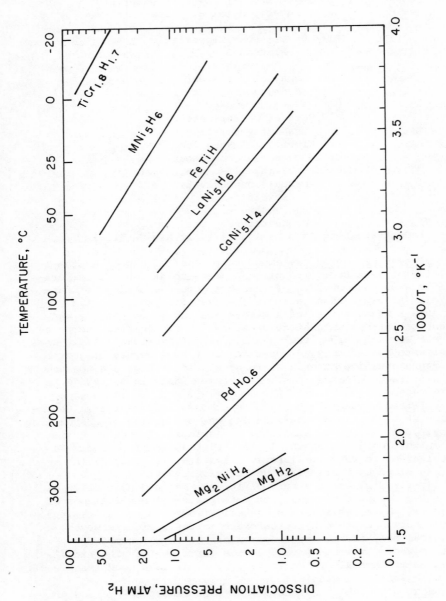

Figure 4. Van't Hoff plots (desorption) for various hydrides

container must effectively also be a heat exchanger. For most applications that require fast cycling, it must be a very effective heat exchanger to make the best use of the hydride.

Hydrogen capacity is a function not only of the crystal structure (available H sites) and base composition of the hydride forming alloy, but also a series of subtle metallurgical factors involved in alloy preparation(2-9). A number of laboratories throughout the world, including ours, are actively seeking new materials with improved capacity. For many of the stationary applications of potential interest to the hydrogen industry, hydrogen capacity (at least on a weight basis) is not a critical property.

Substantial volume changes are associated with hydriding/ dehydriding reactions. A typical example is $LaNi_5$ which expands by about 25% during hydriding and contracts an equal amount on dehydriding(10). Because most of the alloys used for hydrogen storage are brittle, the result is a general breakdown in particle size to a finer powder, i.e., decrepitation. This causes two problems. First, the fine powders must be prevented from being blown out of the container, and this is accomplished by the use of micron sized filters in the exit line. Second, depending on reaction bed design, the fine powders can pack and result in serious bed impedance and expansion problems. Expansion of the packed bed during hydriding can lead to stresses on the container walls that are many times those from the gaseous hydrogen pressure alone (much like the freezing of water in a confined space). We have been able to plastically deform, and even rupture, containers under certain conditions.

Container design is most important. To completely avoid the expansion problem we have developed an encapsulation scheme shown schematically in Figure 5(11,12). The expansion is accommodated by the loosely packed individual capsules, preventing any stress on the main container walls. Each capsule is a thin-walled Al tube containing the hydride and capped on one or both ends with a porous metal filter. Such a design also eliminates the long distance gas impedance problem of a packed bed.

Activation of a hydride forming alloy must be accomplished before it is put into routine service. In some cases (e.g., the AB_5 compounds) the procedure is simply to evacuate the air from the container, pressurize the sample with a suitable overpressure of H_2 at room temperature, and wait a few minutes to a few hours. The pressure required depends strongly on the composition (e.g., 1 atm H_2 is often sufficient for $CaNi_5$, whereas 100-200 atm may be required for MNi_5). The AB compound FeTi requires heating to 300-450°C in H_2 to achieve the start of activation, followed by 2-10 days exposure to high pressure H_2 (>30 atm) at room temperature. For all hydrides the activation process results in a highly cracked structure (Figure 6) with on the order of 0.2-1.0 m^2/g of active surface area. Obviously, from a practical point of view, the easier the activation procedure the

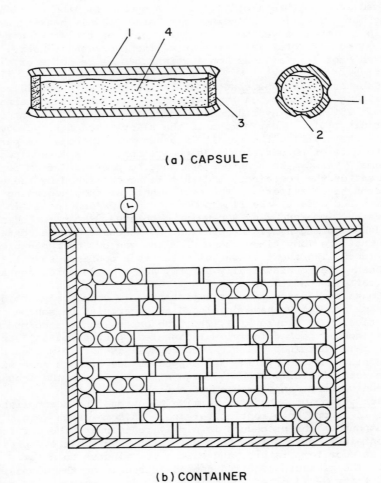

(a) CAPSULE

(b) CONTAINER

Figure 5.   Encapsulation of hydride to prevent expansion and packing problems:
1 = Al capsule; 2 = flute; 3 = porous filter; 4 = hydride.

better. Room temperature activation is especially desirable
because it allows containers (such as aluminum) to be used that
cannot be heated, thus avoiding transfer of the activated hydride.

Once the activated structure of Figure 6 has been achieved,
the kinetics of the practical hydriding/dehydriding reactions are
extremely fast, even at room temperature. In fact, it is often
difficult to measure true kinetics independent of heating and
cooling effects associated with the reactions(9). Thus, from a
practical point of view, the actual kinetics seen depends on the
effectiveness of the heat exchanger design. Charging will occur
as rapidly as the heat of reaction is removed from the hydride
bed and discharging will occur as rapidly as the reaction heat is
put back into the bed.

Tolerance to gaseous impurities in the H_2 used is another
extremely important practical property of hydrides. Gaseous
impurities such as O_2, H_2O, CO, H_2S, etc., tend to "poison" the
active surface of the metal or hydride, resulting in a loss of
kinetics and, ultimately, capacity. This phenomenon is shown
schematically in Figure 7. The purity of the H_2 that can be used
depends on the resistance of the alloy to poisoning. Very
little practical data exists in this area, with most work on
hydrides to date using high purity H_2. We have undertaken major
efforts to understand, and hopefully solve, the poisoning problem.

Chemical stability is an important property that must be
considered for applications involving high temperatures. Undesirable reactions can occur. For example, the desired reaction for
$CaNi_5$ is

$$CaNi_5 + 3H_2 \rightleftarrows CaNi_5H_6 \qquad\qquad (Eq. 3)$$

and, in fact, this reaction works repeatedly and reversibly at
near room temperature. The following disproportionation reaction,
however, is thermodynamically preferred:

$$CaNi_5 + H_2 \rightarrow CaH_2 + 5Ni \qquad\qquad (Eq. 4)$$

At temperatures on the order of 200°C, where diffusion of the
metal atoms becomes significant, the latter reaction tends to
begin and results in a loss of reversible capacity (i.e., irre-
versible Eq. 4 tends to predominate over reversible Eq. 3).
Although most of the intermetallic compounds tend to be thermodynamically unstable relative to disproportionation, their
actual tendency to do so varies markedly from system to system.
For example, $LaNi_5$ is clearly much more resistant to disproportionation than $CaNi_5$. For applications requiring excursions to
high temperatures (see the compressor described later), the
material must be chosen with consideration of the chemical
stability.

Thermal conductivity and specific heat are properties that
should be considered in heat exchanger design of hydride beds.

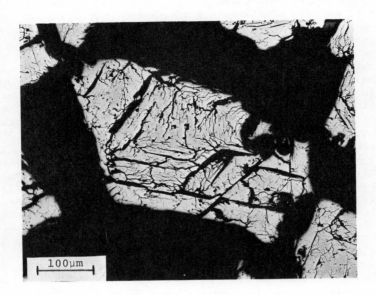

Figure 6. *Metallographic cross section of activated FeTi particle: magnification,*
200×.

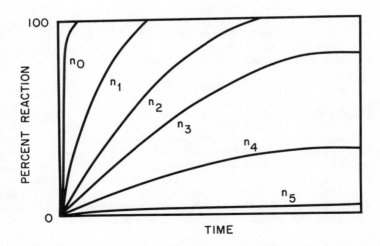

Figure 7. *Schematic reaction curves showing loss of kinetics and capacity as a*
result of cycling with impure H_2.

Specific heats of metals and hydrides are easily determined and
typically fall in the range of 0.1-0.2 cal/g°C. Thermal conductivity is a little more difficult to determine. The conductivity
of the metal or hydride phase is not sufficient; the effective
conductivity of the bed must be determined. This depends on
alloy, particle size, packing, void space, etc. Relatively
little data of an engineering nature is now available and must be
generated for container optimization. Techniques to improve
thermal conductivity of hydride beds are needed. As pointed out
earlier, good heat exchange is the most important factor in rapid
cycling.

The inherent safety of hydrides(13,14) gives them an advantage over compressed gas and liquid hydrogen storage and handling.
The small void space and low pressures involved mean there is
little gaseous H_2 immediately available for catastrophic release
in a tank rupture situation. The endothermic, self limiting,
nature of the desorption reaction also tends to limit the rate of
accidental discharge after rupture. However, some unique care
must be taken with hydride storage. Active hydride powders can
be mildly pyrophoric on sudden exposure to air. For example, the
AB_5 compounds will begin to glow like coal a few minutes after
being suddenly exposed to air in the activated condition. As
discussed earlier, it is extremely important to avoid the expansion problem. Finally, because of the nature of the Van't Hoff
plot, very high pressures can be generated if a charged hydride
is accidentally heated too much. All hydride containers should
have adequate pressure relief devices for fire and other potential
accident situations.

Hydriding alloys represent a new class of engineering
materials requiring the learning of new metallurgical production
factors. Historically, we have been especially active in this
area(2-8). We feel the state of production quality control is
now in a reasonably good condition. As will be discussed later,
we now produce and market a wide variety of hydriding alloys. As
these materials find more and more widespread application, we
must concentrate our efforts even more intensely on potential
cost problems and the long term availability of raw materials.

Families of Rechargeable Metal Hydrides

A brief review of the main classes of available rechargeable
metal hydrides will be given in this section. Most of the
practical (near room temperature) hydride formers are intermetallic compounds consisting of at least one element A that has
a very high affinity for hydrogen and at least one element B that
has a relatively low affinity for hydrogen. Commonly they fall
into three classes of intermetallic compounds: AB, AB_5 and A_2B.
Van't Hoff plots of representative examples are shown in Figure
4.

AB Compounds. The most well known of the AB compounds is
FeTi, developed about 1969 at Brookhaven National Lab(15).   A
hysteresis loop for FeTi is shown in Figure 8.   Note there are
effectively two plateaus representing approximately

$$FeTi + 1/2\ H_2 \rightleftarrows FeTiH \qquad\qquad\qquad (Eq.\ 5)$$

and

$$FeTiH + 1/2\ H_2 \rightleftarrows FeTiH_2 \qquad\qquad (Eq.\ 6)$$

FeTi is the lowest cost room temperature hydride presently available, its principle advantage.   Relative to other hydrides, it
has a few disadvantages: (a) high hysteresis (see Figure 8), (b)
low "poisoning" resistance, at least to O_2(16) and, (c) a heating
requirement for activation.   In addition, it is sensitive to a
number of production variables(4,5,6,7,8).   However, carefully
handled, it is an effective H-storage medium and has been used in
a number of prototype storage tanks.

       Like many of the hydride systems, various partial ternary
substitutions can be made in FeTi, providing changes in plateau
pressure and greatly increasing its versatility(5,17).   Examples
of elements that can be partially substituted into FeTi are Mn,
Cr, Co, Ni and V.   Ni is very effective at lowering the plateau
pressure.   Mn is an especially attractive substitution because
(Fe,Mn)Ti can be activated at room temperature(16), eliminating
the need for heating the container.

       AB5 Compounds. The classic AB_5 hydrogen storage compound is
LaNi5, developed at the Philips Laboratories around 1969(10).   A
25°C hysteresis loop for LaNi5 is shown in Figure 9.   It shows
classic absorption/desorption behavior and has very attractive
hydrogen storage properties: convenient plateau pressures, low
hysteresis, excellent kinetics, easy activation, and relatively
good resistance to poisoning.

       The primary disadvantage of LaNi5 is its high cost.   We have
had a major program aimed at lowering the cost of the AB5 com-
pounds(2).   The philosophy of this program is shown in Figure 10.
By substituting the low cost rare-earth mixture mischmetal(M) for
expensive La, a much lower cost hydrogen storage compound results.
However, MNi5 has impractically high plateau pressures and
hysteresis so that further substitutions of Ca, Cu, Mn, Fe or Al
must be made.   Al is especially potent in lowering the plateau
pressure.   As shown in Figure 11, plateau pressure can be varied
over a wide range to suit the requirements of the intended application.   Some loss of capacity may result, but on the basis of
raw materials cost per unit of hydrogen storage capacity, we have
developed a number of AB5 ternary compounds that are substantially cheaper than LaNi5 (Table III).

       CaNi5(3) is also a useful storage compound where the avail-
able hydrogen pressure is only 1 atmosphere at room temperature
(Figure 4).   CaNi5 and MNi5 form a continuous solid solution so

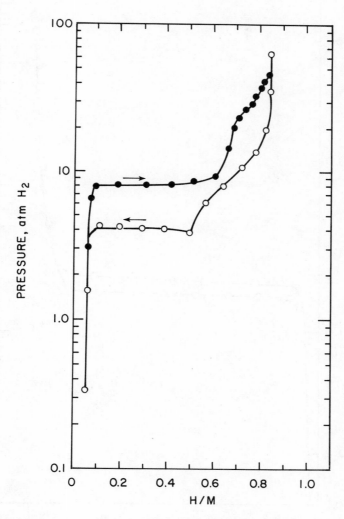

Figure 8.   Absorption and desorption isotherms for FeTi at 16°C (24)

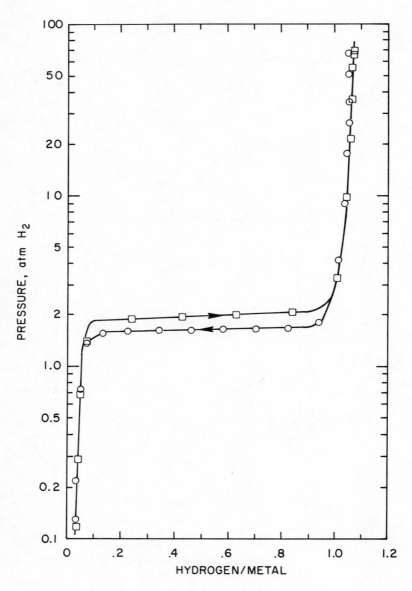

Figure 9.   Absorption and desorption isotherms for LaNi₅ at 25°C

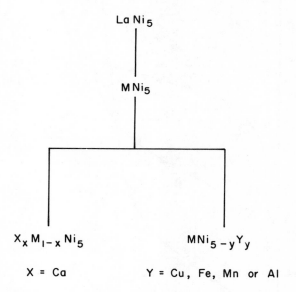

Figure 10. Overall approach to low-cost Ni–rare earth hydrogen storage alloy development (2)

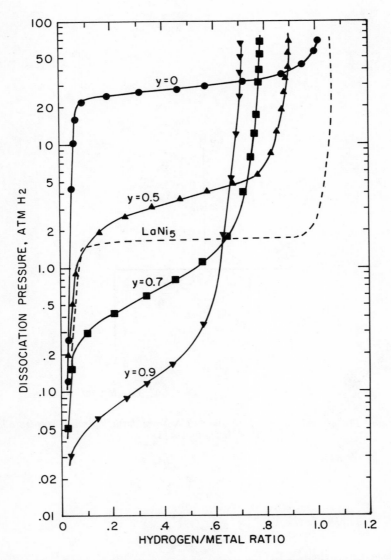

Figure 11.   Effect of Al content y on the 25°C desorption isotherm of As–cast $MNi_{5-y}Al_y$ alloys (2)

Table III. Properties of Selected Room Temperature Hydrides(2)
(Based on 10 Atm-1 Atm Desorption at 25°C, As-Cast Condition)

Alloy	Density, g/cm^3	Dissociation P at H/Metal = 0.5, Atm	$\Delta \dfrac{Hydrogen}{Metal}$	ΔH, Wt.%	Raw Materials Cost*	
					$/Kg Alloy	$/g H-Storage Capacity
Ca$_{0.7}$Mo$_{0.3}$Ni$_5$	7.2	3.8	0.67	1.11	4.98	0.45
MNi$_{3.5}$Cu$_{1.5}$	8.5	8.0	0.69	0.95	5.26	0.55
MNi$_{4.0}$Fe$_{1.0}$	8.3	7.6	0.68	0.95	5.40	0.57
MNi$_{4.5}$Mn$_{0.5}$		4.3	0.82	1.15	5.77	0.50
MNi$_{4.5}$Al$_{0.5}$	8.1	3.7	0.78	1.13	5.92	0.52
LaNi$_5$ (Present)	8.3	1.7	0.97	1.36	17.26	1.27
LaNi$_5$ (Future)	8.3	1.7	0.97	1.36	9.45	0.69
FeTi	6.5	4.2	0.64	1.24	3.08	0.25

* Does not include production costs and profit. Early 1978 costs.

that a complete range of Van't Hoff lines between $CaNi_5$ and MNi_5 on Figure 4 can be achieved(3).

A₂B Compounds. An important A_2B compound is Mg_2Ni, developed at Brookhaven National Laboratory(18). This is one of the most useful examples of the so-called lightweight hydrides (up to 3.8 wt.% hydrogen, Table I). Desorption isotherms of Mg_2Ni are shown in Figure 12. They are extremely flat and, although not shown in Figure 12, have very small hysteresis. Mg_2Ni is a prime candidate for mobile applications (e.g., a fuel carrier for hydrogen powered vehicles). The disadvantage of Mg_2Ni for many stationary applications is the high temperature required for H-desorption, on the order of 300°C. ΔH is also rather high at -15 kcal/mol H_2.

TABLE IV. HY-STOR[+] Alloys Available
Commercially(21)

Trade Name	Alloy
HY-STOR 100 Series (Fe-Base)	
HY-STOR 101	FeTi
103	$(Fe._9Mn._1)Ti$
103	$(Fe._8Ni._2)Ti$
HY-STOR 200 Series (Ni-Base)	
HY-STOR 201	$CaNi_5$
202	$(Ca._7M._3)Ni_5$*
203	$(Ca._2M._8)Ni_5$*
204	MNi_5*
205	$LaNi_5$
206	$(CFM)Ni_5$**
207	$LaNi_4._7Al._3$
208	$MNi_4._5Al._5$*
HY-STOR 300 Series (Mg-Base)	
HY-STOR 301	Mg_2Ni
302	Mg_2Cu

* M = Mischmetal
** CFM = Cerium Free Mischmetal
+ HY-STOR is a trademark of MPD Technology Corporation.

Other Compositions. There are a number of labs throughout the world attempting to develop improved hydrides for hydrogen storage and other applications. Some work is concentrating on Mg-base alloys in an attempt to simultaneously achieve high

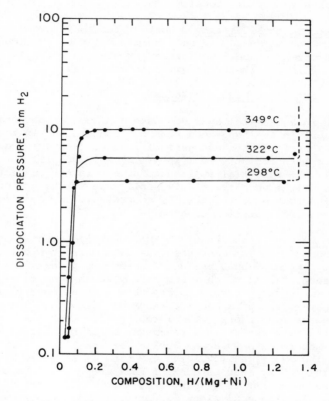

Figure 12.   Desorption isotherms for Mg$_2$Ni (18)

weight percent H and low dissociation temperature. Pure Mg would
be somewhat useful (7.6 wt.% H in MgH_2) but its hydriding kinetics
are impractically slow. Even a few percent Ni addition greatly
improves the hydriding/dehydriding kinetics of Mg(18).

Recently some new AB_2 hydrides have shown promise. Examples
are $ZrFe_2$(19), potentially useful near room temperature, and the
highly unstable $TiCr_2$(20) shown in Figure 4. None of these new
experimental hydrides are yet cost-competitive with FeTi and the
lower cost AB_5 compounds. However, the pace of new developments
in this area is rapidly accelerating.

Commercial Availability. Most of the above hydrogen storage
alloys are now produced and sold commercially by the MPD Technology Corporation under the trademark HY-STOR(21). A list of
the stock alloys marketed is given in Table IV.

Applications in the Hydrogen Industry

Rechargeable metal hydrides have a number of potential
applications in the "hydrogen economy" concept and in the present
industrial sector. Because general surveys of hydride applications have been recently published(1,22,23), we will not try to
review all the potential applications, but rather concentrate on
a few that might be particularly applicable to the hydrogen
industry: storage containers, H_2 compressors, H_2 purification or
separation, and deuterium separation.

Storage and Transportation. Merchant hydrogen is usually
stored and shipped in compressed gas cylinders, typically size K
cylinders, which at 2000 psi contain about 1 pound of H_2. There
are pressure, volume weight, and safety advantages to using
hydrides for this purpose. To see this, let us compare the
storage parameters of a hydride reservoir with conventional
compressed gas cylinders. The large hydride tank that has been
studied the most so far is a rervervoir built by Brookhaven
National Lab for a New Jersey Public Service Electric & Gas Co.
peak shaving experiment(22,24). The tank is shown in Figure 13.
It contains 400 kg (880 pounds) of FeTi and has a reversible
storage capacity for 12 pounds of H_2 when charged to 500 psig
pressure. This is equivalent to about 12 size-K compressed gas
cylinders at 2000 psig. As shown in Table V, there are significant weight and pronounced volume advantages for the hydride
rervervoir over 12 compressed gas cylinders.

Table V. Comparison of BNL PSE&G Hydride Tank with
Compressed Gas Cylinders

Storage Method	H_2 Stored, (lbs)	Total Wt., (lbs)	Total Volume (ft^3)
FeTi Hydride Reservoir	12	1250	5.1
12 Size K Cylinders (2000 psi)	12	1608	18.5

Comparisons of hydride storage to liquid hydrogen should
also be made. Relative to liquid H_2 storage, hydride storage is
energy efficient. The cryogenic liquefaction process for H_2
requires an equivalent of upwards of 33% of the combustion
energy of the H_2 being liquefied. The heats of hydriding reactions are only 10-15% of the combustion energy of the H_2 involved,
and more importantly, represent low cost, low grade "waste" heat
(e.g., room temperature heat). Liquid H_2 requires insulated
tanks; hydrides do not. Unlike hydride storage, liquid H_2
cannot be stored for long periods of time because of boiloff
losses. Incidentally, the capture of boiloff H_2 is another
possible small scale application for metal hydrides within the
liquid portion of the hydrogen industry.

The principal disadvantages of hydride storage and shipment
of H_2 are the cost of the more complex heat exchanger tanks and
the cost of the hydrogen storage alloy itself. The complexity of
the heat exchanger depends on the charging and discharging rates
that may be required. As far as the storage alloy is concerned,
FeTi is presently the cheapest of the room temperature hydrides,
but will cost on the order of $2.20/lb in 1977 dollars(8). Thus,
for the PSE and G tank, the cost of the storage alloy alone is on
the order of $1,936 or about $161/lb of H_2 storage capacity.
Overall hydride container cost studies have never been done, so
it remains to be seen if the additional cost of hydride tanks
over carbon-steel compressed gas cylinders can be economically
justified relative to the hydride reservoirs lower weight and
volume advantages.

H_2 Compressors. Metal hydrides offer an attractive alternative to mechanical compression of H_2 gas for industrial applica-
tions(25). The concept is very simple and is shown in Figure 14.
Although there are essentially no moving parts in a hydride
compressor, it operates by a two-step procedure analogous to a
mechanical compressor and uses temperature increases to increase
the pressure according to the Van't Hoff equation (Eq. 2).
Referring to Figure 14, the "intake" step involves absorption of
H_2 into the bed at low temperature and pressure and the "exhaust"
step the desorption of H_2 at a higher bed temperature and hence
higher pressure. Referring to typical Van't Hoff plots (Figure

CRC Press, Inc.

Figure 13. Large FeTi hydrogen storage tank built at Brookhaven National Laboratories (22)

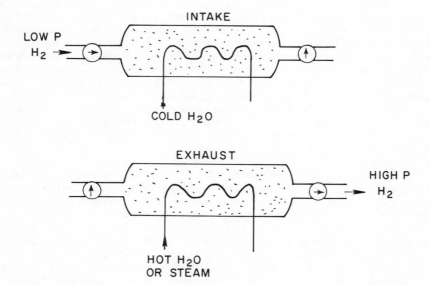

Figure 14. Thermal compression of hydrogen using hydrides

4), it is evident that only modest temperature increases are
required to achieve substantial compression ratios. In most
cases, low grade waste heat sources readily available in many
industrial environments, can be used instead of electric or
fossil fuel power conventionally used. The only moving parts are
check valves.

Hydride compressors offer a number of potential advantages
over mechanical hydrogen compressors(25); lower capital cost,
longer maintenance free life, lower energy cost, (i.e., the use
of low grade "waste" heat) and quieter (vibration-free) operation. Small laboratory prototypes have already been built and
operated(26,27). Ergenics/MPD Corp. and DRI are developing a
larger scale commercial unit(25).

Many of the design principles of a hydride tank also hold
for a hydride compressor. It must be an effective heat exchanger.
The more effective the heat exchange, the shorter is the cycle
time and consequently the smaller the hydride inventory required.
Ideally, the hydride should have a high slope to the Van't Hoff
plot to produce maximum compression with minimum temperature
excursion. Beds of different hydrides can be coupled in series
to provide staged compression and thus achieve very high overall
compression ratios with modest temperatures. The ability to
tailor hydrides, as shown earlier, is very helpful to compressor
optimization for specific applications (i.e., available heat
sources, H_2 input pressure, and desired output pressure).

Purification or Separation. Because hydrides absorb only H_2
and not other gases, they offer the opportunity for purification
or H_2 separation from mixed gas streams. They offer a compli-
mentary technology to pressure swing adsorption techniques. The
basic two-step hydride process is shown in Figure 15. The partial
pressure of H_2 in the H_2-X mixture must be kept above the plateau
pressure during the absorption (separation) stage so that H_2 is
selectively absorbed into the bed. During the thermal discharge
of the bed then, extremely high purity H_2 is desorbed from the
hydride. Ideally, the design should be a circulating system to
prevent impurity gas blanketing around the hydride particles
(i.e., a localized lowering of the H_2 partial pressure to the
plateau pressure as the H_2 is selectively absorbed locally).

Experiments at KFA Jülich have shown that H_2 can be purified
to greater than 99.9999% using a FeTi hydride tank(28). Separation of H_2 from mixtures of 10-50% volume % H_2 in CH_4 has been
demonstrated using Fe_xTiNi_{1-x} alloys(29). The most important key
to the successful application of hydrides for H_2 separation and
purification clearly will be the practical prevention of deactiva-
tion (poisoning) by the impurities present in the H_2.

Deuterium Separation. Most hydrogen storage alloys show
similar absorption/desorption properties for hydrogen and deuterium. Occasionally, however, the hydride and deuteride show

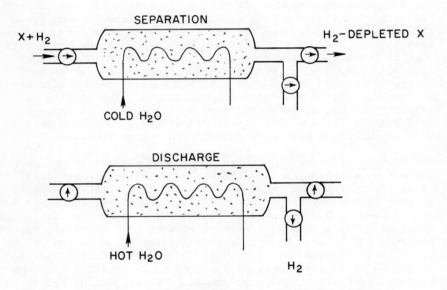

Figure 15.   Hydrogen separation–purification using hydrides

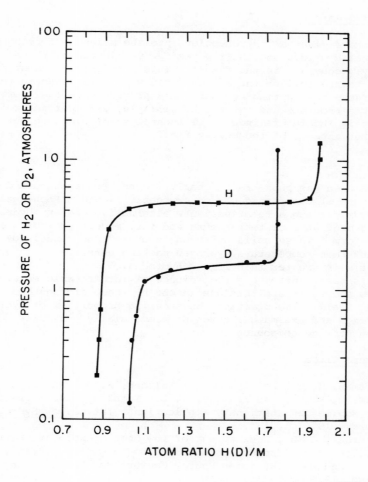

CRC Press, Inc.

Figure 16.   Hydrogen and deuterium absorption isotherms (40°C) for vanadium
(22)

substantially different plateau pressures. An example is V,
shown in Figure 16. This offers a possible technique of frac-
tionally separating deuterium from "natural H_2" (0.015% D).
Deuterium is used in heavy water reactors and will be the fuel of
the future fusion reactor. In addition to V(29), TiNi has shown
promise as a deuterium separation alloy(30).

Concluding Remarks

We have given an introduction into the science of rechargeable metal hydrides and their potential use in the hydrogen
industry, today and in the future. This brief paper cannot do
justice to the full potential and world-wide activities in this
area. These fascinating new materials provide much fuel for the
inventor's thoughts and, we think, will play an important role
not only in the future industrial technology of hydrogen but in
many other aspects of the energy field.

Abstract

Hydrogen has been traditionally stored, transported and used
in the form of compressed gas or cryogenic liquid. Rechargeable
metal hydrides are a relatively new alternative whereby hydrogen
can be stored at room temperature and very modest pressures in
"solid" form. These solid hydrogen "sponges" are capable of
achieving room-temperature hydrogen packing densities that are
substantially greater than liquid hydrogen.
This paper introduces the science of rechargeable metal
hydrides and their applications in the hydrogen storage and
handling field. The practical engineering properties of hydrides
are defined and the main classes of commercially available
hydrogen storage compounds surveyed.

Literature Cited

1. Hoffman, K. C.; Reilly, J. J.; Salzano, F. J.; Waide, C. H.;
   Wiswall, R. N.; and Winsche, W. E., Metal Hydride Storage
   for Mobile and Stationary Applications, Int. J. of Hydrogen
   Energy, 1976, 1, 133.
2. Sandrock, G. D., Development of Low Cost Nickel-Rare Earth
   Hydrides for Hydrogen Storage, "Hydrogen Energy System",
   Proc. 2nd World Hydrogen Energy Conference, Zurich, Pergamon
   Press, 1978 3, 1625.
3. Sandrock, G. D., A New Family of Hydrogen Storage Alloys
   Based on the System Nickel-Mischmetal-Calcium, "Proc. 12th
   IECEC", 1977, I, 951, Washington, DC.
4. Sandrock, G. D., The Metallurgy and Production of Recharge-
   able Hydrides, "Hydrides for Energy Storage", Proc. Inter-
   national Symposium on Hydrides for Energy Storage, Geilo,
   Norway, Pergamon Press, 1977, 353.

5. Sandrock, G. D.; Reilly, J. J.; and Johnson, J. R., Metallurgical Considerations in the Production and Use of FeTi
   Alloys for Hydrogen Storage, "Proc. 11th IECEC", 1976, $\underline{I}$,
   965, Stateline, NV.

6. Sandrock, G. D.; Reilly, J. J.; and Johnson, J. R., Interrelations Between Phase Diagrams and Hydriding Properties
   for Alloys Based on the Intermetallic Compound FeTi, "Applications of Phase Diagrams in Metallurgy and Ceramics",
   National Bureau of Standards SP-496, 1977, 483.

7. Sandrock, G. D., "The Interrelations Among Composition Micro-
   structure, and Hydriding Behavior for Alloys Based on the
   Intermetallic Compound FeTi", International Nickel Co.,
   Suffern, NY, 10901, Final Report for Contract BNL 352410S,
   June 30, 1976.

8. Sandrock, G. D.; and Trozzi, C. J., "Thermodynamic, Economic
   and Metallurgical Studies of Various Techniques for the
   Large Scale Production of Hydriding Grade FeTi and Related
   Compounds", International Nickel Co., Suffern, NY, 10901,
   Final Report for Contract BNL 352410S (Second Year),
   December 30, 1977.

9. Goodell, P. D.; Sandrock, G. D.; and Huston, E. L., "Micro-
   structure and Hydriding Studies of AB_5 Hydrogen Storage
   Compounds", International Nickel Co., Suffern, NY, 10901,
   Final Report for Sandia Contract 13-0524, December 12, 1978,
   in Press.

10. Van Vucht, J. H. N.; Kuijpers, F. A.; and Bruning, H. C. A.
    M., Reversible Room Temperature Absorption of Large Quanti-
    ties of Hydrogen by Intermetallic Compounds, Philips Res.
    Report, 1970. $\underline{25}$, 133.

11. Turillon, P. P.; and Sandrock, G. D., Hydride Container,
    U.S. Patent 4,133,426, January 9, 1979.

12. Turillon, P. P.; and Sandrock, G. D., Hydrogen Storage
    Module, U.S. Patent 4,135,621, January 23, 1979.

13. Lundin, C. E.; and Sullivan, R. W., The Safety Character-
    istics of $LaNi_5$ Hydrides, "Proc. THEME Conf.", University of
    Miami, 1974, S4-36.

14. Lundin, C. E.; and Lynch, F. E., Safety Characteristics of
    FeTi Hydrides, "Proc. 10th IECEC", 1975, Newark, DE, 1386.

15. Reilly, J. J.; and Wiswall, R. H., Formation and Properties
    of Iron Titanium Hydride, Inorganic Chem., 1974 $\underline{13}$, 218.

16. Reilly, J. J.; and Johnson, J. R., Metal Hydride Materials
    Program at BNL - Current Status and Future Plans, "Proc. ERDA
    Contractors' Review Meeting on Chemical Energy Storage and
    Hydrogen Energy Systems", ERDA Report CONF-761134, 1976, 129.

17. Reilly, J. J.; and Johnson, J. R., Titanium Alloy Hydrides
    and Their Applications, "Proc. 1st World Hydrogen Energy
    Conf.", 1976, Miami Beach, International Assoc. of Hydrogen
    Energy, II, 8B-6.

18. Reilly, J. J.; and Wiswall, R. H., The Reaction of Hydrogen
    With Alloys of Magnesium and Nickel and the Formation of
    Mg$_2$NiH$_4$, Inorganic Chem., 1968, 7, 2254.
19. Jacob, I.; and Shaltiel, D., The Influence of Al on the
    Hydrogen Sorption Properties of Intermetallic Compounds,
    "Hydrogen Energy Systems", Proc. 2nd World Hydrogen Energy
    Conf., 1978, Zurich, Pergamon Press, 3, 1689.
20. Johnson, J. R.; and Reilly, J. J., The Reaction of Hydrogen
    With the Low Temperature Form (C15) of TiCr$_2$, Preprint
    BNL-24253, Brookhaven National Lab., February 1978.
21. Anon, "HY-STOR Metal Hydrides - A Revolution in Hydrogen
    Storage Technology", Sales Brochure, MPD Technology Corp.,
    4 William Demarest Pl., Waldwick, NJ 07463.
22. Reilly, J. J., Metal Hydrides as Hydrogen Storage Media
    and Their Applications, "Hydrogen: Its Technology and
    Implications", 1977, CRC Press, II, 13.
23. Lynch, F. E.; and Snape, E., The Role of Metal Hydrides in
    Hydrogen Storage and Utilization, "Hydrogen Energy Systems",
    1978, Zurich, Pergamon Press, 3, 1475.
24. Strickland, G.; Reilly, J. J.; and Wiswall, R. H., An
    Engineering Scale Energy Storage Reservoir of Iron Titanium
    Hydride, "Proc. THEME Conf.", 1974, U. of Miami, S4-9.
25. Lundin, C. E.; Lynch, F. E.; and Snape, E., The Development
    of Metal Hydrides Systems for Hydrogen Compressor Applica-
    tions, Ergenics MPD Corp. and Denver Research Institute,
    DRI Proposal No. MT7904, August 1978.
26. Reilly, J. J.; Holtz, A.; and Wiswall, R. H., A New Labo-
    ratory Gas Circulation Pump for Intermediate Pressures,
    Rev. of Scientific Instruments, 1971, 42, 1485.
27. Van Mal, H. H., A LaNi$_5$-Hydride Thermal Absorption Compressor
    for a Hydrogen Refrigerator, Chemie-Ing.-Techn., 1973, 45,
    80.
28. Klatt, K. H.; Pietz, S.; and Wenzl, H., "Verwendung von
    FeTi-Hydrid Zur Herstellung und Speicherung von höchstrei-
    nem Wasserstoff", Report 770921, Institut für Festkorper-
    forschung, KFA, Jülich, Germany.
29. Cholera, V.; and Gidaspow, D., Hydrogen Separation and Pro-
    duction from Coal-Derived Gases Using Fe$_x$TiNi$_{1-x}$, "Proc. 12th
    IECEC", 1976, Stateline, NV, I, 981.
30. Buchner, H., The Hydrogen/Hydride Energy Concept, "Hydrides
    for Energy Storage", Proc. International Symposium on
    Hydrides for Energy Storage, 1977, Geilo, Norway, Pergamon
    Press, 569.

RECEIVED July 12, 1979.

Closing the Loop for the Sulfur-Iodine Cycle

G. CAPRIOGLIO, K. McCORKLE, and R. SHARP

General Atomic Company, P.O. Box 81608, San Diego, CA 92138

Research on methods of hydrogen production by thermochemical decomposition of water has been carried out in several laboratories with increasing interest during the last decade.

From a thermodynamic point of view, many water-splitting cycles are possible and have been proposed: chemical feasibility, thermal efficiency and engineering and cost considerations have gradually enacted a selection mechanism and only few cycles are actively being studied today.

At General Atomic (GA), attention has been devoted during the last five years to the sulfur-iodine cycle and a program under the joint sponsorship of the U.S. Department of Energy, the Gas Research Institute (previously the American Gas Association) and General Atomic is presently under way. The sulfur-iodine cycle is characterized by the following chemical equations:

$$I_2 + SO_2 + 2H_2O \rightarrow H_2SO_4 + 2HI \tag{1}$$

$$2HI \rightarrow H_2 + I_2 \tag{2}$$

$$H_2SO_4 \rightarrow H_2O + SO_2 + \tfrac{1}{2}O_2 \tag{3}$$

and presents several favorable properties which have been previously reported ([1,2,3,4]).

A brief summary of these properties includes:

a) well characterized chemical reactions involving only fluids

b) heat utilization within a temperature range accessible to heat sources utilizing existing materials technology (specifically, a High-Temperature Gas-Cooled Reactor, HTGR)

0-8412-0522-1/80/47-116-323$05.00
© 1980 American Chemical Society

c)   thermal efficiency of about 50% based on a realistic and
cost conscious flow sheet.

The General Atomic program includes four areas of development:

1.   Chemical Investigations

2.   Materials Investigations

3.   Process Engineering

4.   Bench-Scale Testing

Efforts in the chemistry area are being focused on the I_2-
SO_2-H_2O reaction with particular attention to the separation and
treatment of the two products for subsequent decomposition. Also
included are the studies on the catalytic decomposition of H_2SO_4
and HI. A description of the chemical investigation has been
given by Norman, et al. (4,5).

Materials investigations are directed toward establishing
corrosion resistance to the various process fluids: H_2SO_4 and
its decomposition products and several solutions of HI, H_2O and
I_2 [Trester and Liang, (6)].

The formulation and optimization of a complete flow sheet
is the basis of the process engineering work. Process engineering
and bench-scale studies have been reported by de Graaf, et al.
(7).

The Bench Scale Investigations of the sulfur-iodine cycle
include a system which was planned to study the cycle under continuous operation conditions and a smaller unit, the Closed Loop
Cycle Demonstrator, aimed at a simple demonstration of the feasibility of the cycle in a closed loop using recycled materials.

The Bench Scale System consists of three subunits corresponding to the three basic reactions of the cycle:

-   Subunit I (HI-H_2SO_4 production and separation) supplies
    I_2, H_2O and SO_2 to a reactor where these ingredients are
    mixed and equilibrated, forming two liquid product phases
    which are later separated.

-   Subunit II (H_2SO_4 concentration and decomposition) concen-
    trates and purifies H_2SO_4 followed by its vaporization
    and decomposition.

- Subunit III (HI concentration and decomposition) separates I_2 and H_2O from the HI_x phase by first contacting the de-gassed HI_x with H_3PO_4 followed by a column distillation of HI. The H_3PO_4 is reconcentrated and recycled. The HI is decomposed to H_2 and I_2, the latter being recycled to subunit I.

Closed Loop Cycle Demonstrator

During the last part of August 1978, work began on a smaller unit aimed at an earlier simple demonstration of the cycle in a closed loop using recycled materials. This unit, called the Closed Loop Cycle Demonstrator (CLCD), was not designed to dupli-cate the conditions of the process engineering flowsheet, but rather to demonstrate the feasibility of the cycle through a series of related unit operations. Its schematic diagram is shown in Figure 1.

H_2O, I_2 and SO_2 are fed into a main reaction vessel (R-1), and continuously recirculated, by means of a pump P-1, in a loop where temperature control can be obtained through the cooler (or heater) C-1. Some of the products are intermittently fed to the liquid-liquid separator R-2 where the two acid phases are separa-ted. The upper H_2SO_4 phase is purified by boiling in R-3, and then decomposed in a quartz cracker containing Fe_2O_3 catalyst. The product gas is first cooled in a condenser R-5, the condensed phase (unreacted H_2SO_4) recycled to R-3, and the gas phase (SO_2 and O_2) fed to the purification system (TR-1 and TR-2 are two CO_2 acetone traps in parallel). The lower phase from R-2 (HI-I_2-H_2O containing SO_2) goes to an off-gas reactor R-6 where, under vacuum and at 370°K, SO_2 is extracted and trapped in the liquid nitrogen traps TR-3 and TR-4. Again, intermittently, the degassed lower phase is sent to a cracker R-7 where some of the HI is thermally decomposed in H_2 and I_2. The product gases are first cooled in C-4 and then fed to a series of separators and condensers (R-8 and C-5 and 6) where H_2O, HI and I_2 are separated and sent for recycle to the main reaction vessel R-1. The H_2 gas is purified in the liquid nitrogen traps TR-5 and 6 before metering and collecting.

Construction and assembly of the CLCD took place between October and December 1978. Glass was used throughout the system, with the exception of the high temperature portions R-4/C-3 and R-7/C-4, the two cracker-cooler units which were built in quartz. Figure 2 shows the assembled CLCD.

CLCD Operation

During the first weeks of January, operation of the closed loop cycle demonstrator took place. As a result of the experience

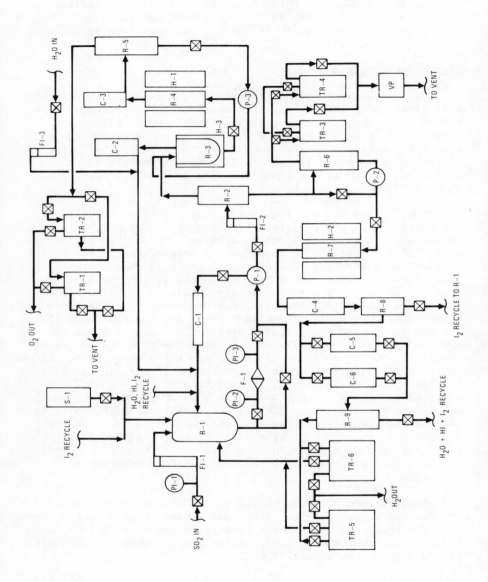

COMPONENT DESIGNATION

C-1	PRIME REACTION RECIRCULATING PRODUCT COOLER
C-2	UPPER PHASE OFF GAS COOLER
C-3	H_2SO_4 POST–CRACK PRODUCT COOLER
C-4	HI POST–CRACK PRODUCT COOLER
C-5	HYDROGEN PURIFICATION LOOP PRE-COOLER
C-6	HYDROGEN PURIFICATION LOOP PRE-COOLER
F-1	PRIME REACTION MAIN–LOOP FILTER
FI-1	SO_2 INPUT FLOW INDICATOR
FI-2	MAIN LOOP OUTPUT FLOW INDICATOR
FI-3	WATER SUPPLY INPUT FLOW INDICATOR
H-1	H_2SO_4 CRACKING FURNACE
H-2	HI DECOMPOSITION FURNACE
H-3	UPPER PHASE PURIFICATION HEATER
P-1	MAIN LOOP CIRCULATING PUMP
P-2	LOWER PHASE CIRCULATING AND FEED PUMP
P-3	H_2SO_4 RECYCLE PUMP
R-1	MAIN REACTION VESSEL
R-2	MAIN REACTION PRODUCT LIQUID–LIQUID SEPARATOR
R-3	H_2SO_4 PURIFICATION BOILER
R-4	H_2SO_4 DECOMPOSITION REACTOR
R-5	UPPER PHASE POST–DECOMPOSITION GAS–LIQUID SEPARATOR
R-6	LOWER PHASE OFF–GAS REACTOR
R-7	LOWER PHASE DECOMPOSITION REACTOR
R-8	LOWER PHASE POST–DECOMPOSITION GAS–LIQUID SEPARATOR
R-9	HYDROGEN PURIFICATION LOOP GAS–LIQUID SEPARATOR
S-1	IODINE SUPPLY VESSEL
TR-1	OXYGEN PURIFICATION LOOP CO_2 – ACETONE TRAP
TR-2	OXYGEN PURIFICATION LOOP CO_2 – ACETONE TRAP
TR-3	LOWER PHASE SO_2 OFF–GAS LN_2 TRAP
TR-4	LOWER PHASE SO_2 OFF–GAS LN_2 TRAP
TR-5	HYDROGEN PURIFICATION LOOP LN_2 TRAP
VP	LOWER PHASE OFF–GAS SYSTEM VACUUM PUMP

Figure 1. Schematic of a CLCD

Figure 2.   Assembled CLCD

gained during testing of the Bench Scale Subunit I, no operational problems were encountered. All valves, joints, pumps and temperature and flow controls worked as designed. For simplicity of operation, it was found convenient to add a small displacement pump feeding the HI cracker. This allowed operation of the degassing P-2/R-6 loop independently from operation of the cracker R-7.

Operation of the CLCD started by introducing 3 Kg of iodine and 1 Kg of water at room temperature in the main reaction vessel R-1. SO_2 was then bubbled through at a rate of 3 liters/minute until there was evidence of formation of two separate liquid phases (20 minutes). The prime reaction products were kept circulating through the filter F-1 and the cooler C-1 by means of the pump P-1. Since the temperature had risen to only 45°C, no cooling was necessary. Intermittently, the liquid was fed to the phase separator R-2, allowed to rest 5-10 minutes to complete the separation and then the upper phase sent to the H_2SO_4 purification boiler R-3 and the lower phase to the degasser R-6.

The H_2SO_4 was concentrated from an initial 40% to approximately 95% and all the water and traces of I_2 and HI cooled in C-2 and recycled into the main loop. The concentrated, purified H_2SO_4 was fed to the cracker R-4, filled with pellets of Fe_2O_3 catalyst and kept at 850°C, at a rate of ~5 cc/minute. The product gases were condensed in C-3, recycling the uncracked H_2SO_4 to the boiler and the resulting SO_2-O_2 mixture directly sent to the main loop for recycle or to the trap system for separation and analysis. The lower phase was degassed (SO_2 removal) under vacuum in R-6 for 10 minutes before it was sent to the HI cracking step.

Since the CLCD was designed, for simplicity, without any HI purification step, the lower phase (HI-I_2-H_2O) was directly fed to the HI cracker and no catalyst was used.

The feeding rate was ~4 cc/minute and the temperature kept at 900°C. Before operating the HI cracker, the system was flushed with helium.

The product gases from the cracker R-7 were cooled and all condensed phases (H_2O, I_2 and unreacted HI) recycled to the main loop. The cooled gas was purified in the liquid nitrogen traps TR-5 and TR-6 and the product H_2 collected in a graduated cylinder.

Operation of the loop was accomplished subsequently in a complete recycle mode: the main reaction products were formed by reacting the recycled I_2 from the HI decomposition system with the SO_2-O_2 mixture obtained from the H_2SO_4 cracker. No differ ences from operation with pure SO_2 were observed.

During operation of the loop, small quantities of sulfur were observed in the recycled liquids from the HI cracking coolers and small amounts of H_2S were collected in the H_2 purification traps. This was due to incomplete separation of the sulfur containing species (SO_2 and H_2SO_4) from the lower phase prior to decomposition. The lower phase concentration and purification step (H_3PO_4 treatment), which is an integral part of the cycle, will eliminate this problem. This step will be tested in the Bench Scale Unit.

Complete operation of the CLCD confirmed the feasibility of of the GA water-splitting cycle and provided the laboratory personnel with information useful for the detailed construction and operation of the Bench Scale Unit.

Conclusion

The sulfur-iodine thermochemical water-splitting cycle continues to represent a promising approach to the production of hydrogen utilizing an entirely thermal energy source. Another significant step toward its viability has been taken by the successful closure of the cycle in this laboratory demonstration.

Acknowledgement

The work described in this paper was conducted under sponsorship of the U.S. Department of Energy (Contract No. EY-76-C-03-0167, P.A. No. 63), The Gas Research Institute (previously American Gas Association) and General Atomic Company; previous work was also sponsored by Northeast Utilities Service Company and Southern California Edison Company.

Abstract

The sulfur-iodine thermochemical water-splitting cycle is simply described by the following three chemical equations:

$$I_2 + SO_2 + 2H_2O \rightarrow H_2SO_4 + 2HI \qquad (1)$$

$$2HI \rightarrow H_2 + I_2 \qquad (2)$$

$$H_2SO_4 \rightarrow H_2O + SO_2 + \tfrac{1}{2}O_2$$

The cycle has been under study at General Atomic during the last five years under the joint sponsorship of the U.S. Department of Energy, the Gas Research Institute (previously the American Gas Association) and General Atomic.

In 1977, the design and construction of a bench-scale unit

was started. The main objective of the unit is the study of the cycle under continuous flow conditions by modeling the main solution reaction (1), product separation and concentration and decomposition of HI and H_2SO_4.

As part of the bench–scale testing program, a separate, smaller unit, called the Closed Loop Cycle Demonstrator (CLCD), has been designed, assembled and operated between September 1978 and January 1979. The CLCD consists of a main reaction vessel where HI and H_2SO_4 solutions are produced from I_2, SO_2 and H_2O, a phase separation and two loops where HI and H_2SO_4 are thermally decomposed feeding the recycled materials back to the main reaction vessel. The operation of the CLCD has confirmed the feasibility of the GA water-splitting cycle and has provided useful information for the construction and operation of the larger Bench–Scale Unit.

Literature Cited

1. Russell, J. L.; McCorkle, K. H.; Norman, J. H.; Porter, J. T.; Roemer, T. S.; Schuster, J. R.; Sharp, R. S.; Water Splitting – A Progress Report," in Proceedings of First World Hydrogen Energy Conference, Miami Beach, Florida, March 1–3, 1976, p. 1A–105.

2. Russell, J. L.; McCorkle, K. H.; Norman, J. H.; Schuster, J. R.; Trester, P. W., "Development of Thermochemical Water Splitting for Hydrogen Production at General Atomic Company," General Atomic Report GA–A14050, September 30, 1976.

3. Schuster, J. R.; Russell, J. L.; Norman, J. H.; Ohno, T.; Trester, P. W., "Status of Thermochemical Water Splitting Development at General Atomic," General Atomic Report GA–A14666, October 1977.

4. Norman, J. H.; Mysels, K. J.; O'Keefe, D. R.; Stowell, S. A.; Williamson, D. G., "Water Splitting – The Chemistry of the I_2–SO_2–H_2O Reaction and the Processing of H_2SO_4 and HI Products," General Atomic Report GA–A14746, December 1977.

5. Norman, J. H.; Mysels, K. J.; O'Keefe, D. R.; Stowell, S. A.; Williamson, D. G., "Chemical Studies on the General Atomic Sulfur–Iodine Thermochemical Water-Splitting Cycle," paper presented at the Second World Hydrogen Energy Conference, Zurich, Switzerland, August 1978.

6. Trester, P. W.; Liang, S. S., "Material Corrosion Investigations for the General Atomic Sulfur–Iodine Thermochemical Water-Splitting Cycle," Second World Hydrogen Energy Conference Proceedings, Zurich, Switzerland, August 1978.

7. de Graaf, J. D.; McCorkle, K. H.; Norman, J. H.; Sharp, R.;
    Webb, G. B.; Ohno, T., "Engineering and Bench-Scale Studies
    on the General Atomic Sulfur-Iodine Thermochemical Water-
    Splitting Cycle," paper presented at the Second World Hydrogen
    Energy Conference, Zurich, Switzerland, August 1978.

RECEIVED July 12, 1979.

Hydrogen from Fuel Desulfurization

M. E. D. RAYMONT

Director of Research & Development, Sulphur Development Institute of Canada, Box 9505, Calgary, Alberta, Canada T2P 2W6

Almost all combustible fuels of commercial significance to man contain at least some sulphur. For many years, low sulphur fuels were able to supply much of our need for energy but in the last two decades as energy demand has surged, many new sources of high sulphur fuels have been developed including Canadian sour gas, U.S. Deep Smackover gas, sour crudes from the Middle East, and even Canadian tar sands deposits containing 4.5 wt.% of sulphur. Future energy sources such as heavy oils, tar sands, oil shales and coals for liquifaction and gasifaction generally contain significant levels of sulphur (often 2-5 wt.%). Coincident with burgeoning energy requirements, strict air pollution regulations have necessitated the removal of sulphur values from fuels. With the exception of coal (and this situation may change in the future) most fuels are desulphurized prior to combustion to eliminate or minimize sulphur dioxide formation.

In the case of gaseous fuels, almost all of the sulphur naturally present will exist as hydrogen sulphide in concentrations ranging from a few parts per million up to 50% or more, although the latter concentrations are somewhat unusual. Desulphurization of gas streams involves solvent scrubbing to separate the raw gas into a cleaned hydrocarbon stream and an acid gas fraction containing the hydrogen sulphide. In liquid fuels, the sulphur is present in many different chemical forms primarily combined in organic molecules such as sulphides, disulphides and mercaptans. To remove it from the liquid fuel, the sulphur values are converted to hydrogen sulphide by reaction with hydrogen. At present solid fuels (coal) are usually combusted prior to removal of sulphur and the resulting sulphur dioxide is scrubbed from the flue gas producing waste sludge or by-product acid. However, most coal liquifaction and gasification processes favor removal of sulphur

0-8412-0522-1/80/47-116-333$05.00
© 1980 American Chemical Society

values as hydrogen sulphide prior to combustion. In addition,
technologies are being developed for the pre-combustion
conversion of sulphur compounds in coal to hydrogen sulphide
to provide a clean solid fuel.

It is therefore clear that the combination of the growing
use of high sulphur fuels and stricter air quality regulations
have resulted in the production of large volumes of hydrogen
sulphide as a by-product of fuel desulphurization. In addition,
substantial quantities of hydrogen, generated from fuels,
are used in the desulphurization process and converted to
hydrogen sulphide. Involuntary hydrogen sulphide generation
has grown rapidly in the last decade averaging more than
10% compounded yearly growth in North America. Because of the
toxic nature of hydrogen sulphide, these streams must be
processed, and at present almost all of this hydrogen sulphide
is converted to elemental sulphur by the Claus Process.

The Claus Process has been used successfully for many
years and currently some 10 million tonnes ($\sim 11 \times 10^6$ tons)
per year of sulphur are produced in North America from
hydrogen sulphide using this technique. This represents an
annual throughput of more than 7 billion cubic metres (250 x
10^9 SCF) of hydrogen sulphide. Typically operating at 94 – 98%
efficiency, the basic Claus Process is relatively efficient
at recovering sulphur, although recent air pollution legislation has required the use of tail gas treatment units to
increase recovery to 99% or better (99.9% + in new plants
in the U.S. and Japan). Unfortunately, however, this process
wastes the hydrogen component of the hydrogen sulphide by
oxidizing it to water which is discharged to the atmosphere
(Fig. 1). Recently, work by this author (1, 2, 3) and groups
in the U.S. and Japan (4-9) has shown that it is technically
possible to recover both sulphur and hydrogen from hydrogen
sulphide. If such a process is economically viable, the
commercial significance is obviously very great.

The Decomposition Reaction

The conventional Claus reaction involves the partial
oxidation of hydrogen sulphide to sulphur and water. In
practice the reaction is generally viewed as a two step
sequence involving a highly exothermic oxidation step followed
by a less exothermic redox reaction but the overall reaction
can be expressed as

$$\Delta H \approx - 650 \text{ kj}$$

$$3H_2S + 3/2O_2 \quad \rightleftharpoons \quad 3/x\ S_x + 3H_2O \qquad (1)$$

By contrast, the decomposition reaction involves an endothermic
dissociation into hydrogen and sulphur,

$$\Delta H \approx \ + 160 \text{ kj}$$

$$2H_2S \quad \xrightleftharpoons{\hspace{2cm}} \quad 2H_2 + 2/x \ Sx \qquad (2)$$

The exothermic Claus reaction is self-sustaining and a net
producer of energy as well as a single useful product,
sulphur. By contrast the endothermic decomposition reaction
must be provided with energy to drive it forward but it
produces two useful products.

Fundamental thermodynamic and kinetic studies of the
decomposition reaction (1) have confirmed that hydrogen
sulphide is a stable sulphide and that the dissociation is
thermodynamically unfavorable below 1800°K. Nevertheless,
some decomposition does, of course, occur below these temperatures and equilibrium hydrogen yields range from less than
1% at 750°K through about 5% at 1000°K to almost 30% at
1400°K. [These values are based on equilibrium product
calculations which considered all possible sulphur/hydrogen
species which could be present at equilibrium including
various sulphur vapor species (S_1 S_2 to S_8), and sulphanes
(H_2Sx) as well as H_2S, H_2 and S_2. The values which are
higher than for the simple system (equation 2) are confirmed
experimentally (1).] These yields under such extreme conditions are not, however, a very encouraging basis on which to
plan a commercial process. In addition, the uncatalyzed
kinetics are not particularly promising; the reaction is
sufficiently slow that below 1000°K it takes fully a minute
to reach equilibrium and even at 1200°K the reaction time is
about 15 secs (1).

Evidently, the uncatalyzed direct thermal decomposition
is simply too unattractive from both a thermodynamic and
kinetic viewpoint to be seriously considered for a practical
process. Work has therefore focussed on methods for increasing the reaction yields under more moderate process conditions.
Unfortunately, thermodynamics cannot be violated but thermodynamic limitations can be sidestepped, and kinetic limitations can be overcome using catalysts. Four principle types
of techniques are being investigated for improved yields
including upset equilibrium systems, closed cycle loops, open
cycle loops, and electrochemical methods. The advantages and
disadvantages of these will now be discussed more fully.

Decomposition Techniques

a)   Upset Equilibrium. If one or more products are
continuously removed from a reaction system then equilibrium
conditions are not satisfied and so unfavorable thermodynamics
can be circumvented. In the case of hydrogen sulphide
dissociation either product (or both) could be removed thus
minimizing the reverse reaction and drawing the process to

favor sulphur and hydrogen formation.

In order to remove hydrogen, a diffusion membrane can
be used. Molecular sieves would seem to be ideal but their
degradation at even moderate temperatures effectively eliminates their use. Palladium or platinum metal alloy membranes
have been used commercially for the purification of hydrogen
(10) and were also found effective in catalysing the decomposition reaction (1). Provided these membranes are not structur-
ally damaged by the inevitable sulphidation, they could act
as both heterogeneous catalysts and selectively permeable
membranes for the product hydrogen. At elevated temperatures
necessary for the reaction, many metals are permeated by
hydrogen and thus other lower cost metal membranes may be
useful. In addition, porous glasses or ceramics should be
viable membrane materials. Recent work in Japan (11) has
shown that porous Vycor glass can be used to enhance hydrogen
yields although kinetic limitations dropped yields to about
half the expected values.

Clearly the decomposition rate must be fast enough to
provide sufficient products for separation. The uncatalyzed
reaction is slow but various metals, notably certain transition metal sulphides, have been found to be effective in
catalyzing the dissociation. Cobalt and molybdenum sulphides
have been found to be especially active as have the sulphides
of ruthenium (9), cobalt/molybdenum mixtures (1), nickel
/tungsten mixtures (1), platinum (1), and iron. It appears
that the activity of these catalysts is strongly dependent on
the stoichiometry of the metal sulphides and that in general
catalytic activity increases when the metal/sulphur atom
ratio is near unity or greater (e.g. $Co_{0.9}S$, FeS). Higher
sulphides (e.g. FeS_2) appear to be almost ineffective as
catalysts. This observation is consistent with suggested
mechanisms (1). Although optimization of a suitable catalyst
remains to be accomplished, there is little doubt that the
decomposition reaction can be brought about rapidly under
moderate process conditions (800 - 1050°K). Nevertheless,
diffusion separation has certain intrinsic limitations.

The yield increases produced by diffusion separation
are ultimately limited by the Graham diffusion law limit
which depends on the molecular weights of the gases to be
separated. For the H_2/H_2S system the maximum separation
(one pass) is given by

$$S_{max} = \sqrt{M_{H_2S}/M_{H_2}} \tag{3}$$
$$= \sqrt{34/2} = 4.1$$

Maximum enriched gas hydrogen yields (in %) are given by

$$Y_{max} = 100 \ \frac{S_{max} \ Y_{eq.}}{100 + Y_{eq.}(S_{max}-1)}$$

Where $Y_{eq.}$ is the equilibrium hydrogen yield at the process conditions.

$$\therefore \quad Y_{max} = 100 \quad \frac{4.1 \, Y_{eq.}}{100 + 3.1 \, Y_{eq.}}$$

Thus diffusion separation is effective in increasing low yield concentrations of product but as the equilibrium yield rises the efficiency of enrichment is reduced. One further drawback arises if the hydrogen sulphide/sulphur stream is recycled. Eventually, if the product sulphur is not also removed, the partial pressure of the sulphur will be so great as to depress the forward reaction even under upset equilibrium conditions.

As an alternative (or adjunct to) hydrogen removal for equilibrium displacement, the sulphur could be removed. A process involving catalytic decomposition followed by sulphur condensation and recycling has been developed at the laboratory level (9). However, the yields per pass are very low and so many cycles involving heating to about 1000°K and cooling to 300°K are necessary to achieve reasonable yields that the process does not have practical applications. Nevertheless, the idea of removing sulphur is appealing and other systems involving removal of sulphur and hydrogen are currently being investigated.

Upset equilibrium systems do increase hydrogen yields but, as yet, kinetic and/or mass action effects limit the viability of such processes.

   b)   Closed Cycle Loops. By breaking a single unfavorable reacton into a number of more favorable steps, higher yields can often be obtained under more moderate conditions. For hydrogen sulphide a generalized two step closed loop is shown below.

$$H_2S + X \longrightarrow XS + H_2\uparrow$$
$$XS \longrightarrow X + (S) \tag{5}$$

X is simply a sulphur "carrier" which can be regenerated. Most work in this area has focussed on the use of sulphides of variable stoichiometry especially those of the transition metals iron, cobalt and nickel. A typical reaction system is shown below.

$$Co_3S_4 + 2H_2S \longrightarrow 3CoS_2 + 2H_2 \tag{6}$$
$$3CoS_2 \xrightarrow{H} Co_3S_4 + S_2 \tag{7}$$

The hydrogen liberation reactions (eq. 6) are favored at lower temperatures and would typically be carried out at 550 - 800°K whereas the second regeneration step (eq. 7) is favored at higher temperatures in the 950 - 1200°K range.

Substantially complete reaction of hydrogen sulphide with iron sulphide has been claimed (12) and the lower sulphide was effectively regenerated by heating in a carbon dioxide stream to remove the sulphur vapor. The cycle was apparently repeated with similar results. Other work using iron, cobalt and molybdenum achieved less complete conversions (60 - 70%) and found the initial reaction to be fairly slow (13). On regeneration, the sulphides appeared to lose some surface area and thus their total conversion capacity for a second pass was reduced. This seemed particularly true with the iron sulphide system. Japanese investigations have however found good retention of surface area, and therefore activity, with calcined natural pyrite, but observed poor repeat yields from the Ni_3S_2/NiS system due to sintering or melting and consequent surface area loss.

Thermodynamic calculations suggest that the maximum conversion efficiencies for the first step will be of the order of 90% for most transition metal sulphides (4). However, the kinetics of this step may not be attractive due to the slow diffusion rate of hydrogen sulphide through the surface-formed dense sulphide film. To overcome this problem, the use of molten metals such as lead has been suggested (4,6). In addition to eliminating the solid state diffusion problem, the use of molten lead can theoretically produce 99.8% conversion to hydrogen at 775°K and one atmosphere. However, again the kinetics of the first step are slow, although they can be improved by the addition of small amounts of nickel or copper to the melt. In Japan, other closed loop systems using transition metal salts instead of metal sulphides are also under investigation. For example, a magnetite/hydrochloric acid cycle has been proposed (14).

$$Fe_3O_4 + 6HCl + H_2S \longrightarrow 3FeCl_2 + 4H_2O + S$$

$$3FeCl_2 + 4H_2O \text{ (g)} \longrightarrow Fe_3O_4 + 6HCl + H_2$$

While closed loops involving metal sulphides or other salts appear attractive, kinetic limitations during the hydrogen generation step must first be solved before a practical process can be realized; surface area considerations are also important. The other disadvantage of these processes is that the amount of material which must be used to convert even small volumes of hydrogen sulphide is substantial. If a batch process is used, these factors may make the closed loop technique uneconomic. However, the possibility of using a fluidized bed system justifies continued investigation of these techniques.

c)   Open Cycle Loops.   In open loop systems, a favorable reaction is "combined" with the unfavorable reaction in order to enhance product yields.   Many open loops involving hydrogen sulphide are possible.   Several have been suggested which use the oxidation of carbon or carbon compounds as the necessary source of negative free energy (1,3).   A typical system is shown below.

$$2H_2S + 2CO \longrightarrow 2H_2 + 2COS \qquad (8)$$

$$2COS + SO_2 \longrightarrow 2CO_2 + 3/2 \ S_2 \qquad (9)$$

$$1/2 \ S_2 + O_2 \longrightarrow SO_2 \qquad (10)$$

The overall reaction is

$$2H_2S + 2CO + O_2 \longrightarrow 2H_2 + S_2 + 2CO_2 \qquad (11)$$

Each of the individual reactions (eq. 8 through 10) have been studied and can be brought about giving good yields (eq. 8 refs. 8,15; eq. 9 ref. 16; eq. 10 well tried and highly exothermic).   However, a major problem with such a scheme involves the separation of reaction products especially in reaction 8.   Both imperfect separation and side reactions (e.g. to form CS_2) give rise to unwanted impurities in the product streams.   These impurities tend to be toxic and/or pollutants and further refining of the product streams and waste disposal would be complex and expensive.   The same criticisms can be applied to a similar open loop involving hydrocarbons, for example, methane.

$$2H_2S + CH_4 \longrightarrow CS_2 + 4H_2 \qquad (12)$$

$$CS_2 + SO_2 \longrightarrow CO_2 + 3/2 \ S_2 \qquad (13)$$

$$1/2 \ S_2 + O_2 \longrightarrow SO_2 \qquad (14)$$

The overall reaction is

$$2H_2S + CH_4 + O_2 \longrightarrow 4H_2 + S_2 + CO_2 \qquad (15)$$

In addition to incomplete reaction and separation problems, these techniques are likely to be expensive because of their multi-step nature requiring separate process units for each phase of the reaction sequence.

In spite of these drawbacks open loop systems are
receiving more attention. An important criterion when
assessing possible open loops is that the final reaction
products of the driving reaction must, at the least, be non-
polluting, and for preference they should be themselves
useful. Clearly the driving reacton must have a negative
free energy and in "combining" the reactions a minimum
number of reaction sequences should be used. Using these
guidelines a number of possible loops can be constructed and
several are now being examined in more detail.

       d)   Electrochemical Methods. In addition to heat and/or
"chemical" energy, the use of electrical energy has been
considered as a method for decomposing hydrogen sulphide.
Pure hydrogen sulphide can be electrolysed in the liquid
state or alternatively, electrolysis can be carried out in
aqueous solution or in a sulphide salt melt.
The basic cell relations to be achieved are

$$\text{Cathode reaction } H_2S + 2e^- \longrightarrow S^{2-} + H_2$$

$$\text{Anode reaction } \quad S^{2-} \longrightarrow 2e^- + 1/2\ S_2$$

$$\text{Overall reaction } \quad H_2S \longrightarrow H_2 + 1/2\ S_2$$

The cathode reaction involves liberation of hydrogen at
the electrode in a reduction step producing S^{2-} ions. A
medium for S^{2-} ion conduction is necessary so that the
oxidizing anode reaction to produce elemental sulphur can
proceed.
       The aqueous electrolysis has been patented (17) although
the author is unaware of any current development work or
commercial applications. The theoretical decomposition
voltage is only about 0.18v at 300°K (at partial pressure of
1 atm. for H_2 and H_2S) although, of course, in practice a
higher voltage would be required to overcome electrode
over-voltages. Under ambient operating conditions the product
sulphur would be solid and some periodic extraction and
drying system would be required. In addition, some pressurization of the cell may be necessary to prevent release of
hydrogen sulphide. Overall the concept appears attractive
although electrode processes would have to be carefully
controlled to prevent further oxidation of sulphur at the
anode thus giving rise to unwanted products and consuming
excess power. For liquified hydrogen sulphide, a refrigerated
system is possible but a pressurized cell operating at or
above ambient temperature would be far more practical.
Because of the toxicity of hydrogen sulphide such cells
would have to be very carefully constructed to maintain
sufficient pressure while permitting easy periodic removal

of the product sulphur which would be in the solid form.
The third alternative involves the high temperature electrolysis of molten sulphide salts. At 750°K the decompositon
voltage is only 0.22 volts, and electrode over-voltages are
likely to be small. Molten salts such as sodium or potassium
sulphide could be used though probably a eutectic mix with
some other sulphides would be necessary in order to achieve
a practical melt. Cell construction materials should be
similar to those being developed for high temperature sodium/
sulphur batteries (graphite electrodes, alumina cells). An
advantage of the high temperature electrolysis is the fact
that both products would be gases at the operating conditions
thus facilitating their removal from the cell. In addition,
the kinetic limitations and over-voltages at the electrodes
should be minimized. Against this, the cell temperature has
to be maintained above 725°K and cell materials may be a
problem.

Electrochemical methods are being investigated because
of their basic simplicity and efficient separation of the
products. Electrical power, of course, is generally more
expensive than other forms of energy but the flexibility of
unit size and the "cleanliness" of such processes make them
potentially attractive especially for small installations.

In summary, several processes warrant serious consideration as practical techniques for the decomposition of
hydrogen sulphide. However, until their relative efficiencies
and limitations have been defined, the optimum process
cannot be identified. Further work on a variety of these
options is warranted.

Commercial and Economic Aspects

At present almost all the hydrogen sulphide involuntarily
produced by fuel desulphurization is processed in Claus
plants. Thus, if hydrogen sulphide decomposition is to
become a commercially viable process, its overall economics
must compare favorably with those of the Claus.

Most Claus plants form an integrated part of a refinery
or natural gas plant and are considered to be a necessary
"evil" since hydrogen sulphide must be treated. Several
years ago, overall Claus plant hydrogen sulphide processing
costs were moderate and could almost invariably be easily
recovered by sale of the product sulphur and recovery of
useful steam from the front-end furnace. However, stringent
air polluton regulations covering Claus emissions and capri-
cious sulphur markets and prices have significantly altered
this situation. Recent estimates by the U.S. Environmental
Protection Agency (18, 19) and the authors suggest that current
sulphur production costs, including tail gas treatment
costs, exceed $50/tonne for small plants (10 - 50 tonnes/day),

and only drop below $20/tonne for very large plants (over 500 tonnes/day). Thus, with sulphur priced typically at $40 - 60/tonne, sulphur production from hydrogen sulphide is no longer necessarily a profitable operation, especially in small plants at refineries and small gas fields.

Against this rather bleak economic backdrop, the commercial potential for the decomposition process appears quite attractive. Obviously, since a single optimized process has yet to be identified, the detailed economics of the decomposition technique are impossible to estimate. Nevertheless, some instructive cost comparisons can be made.

The major difference between the Claus and decomposition reactions is that the former produces heat and one saleable product while the latter requires heat but produces two saleable products. Since no capital costs can be accurately estimated, the gross assumption is made that for the decomposition process these costs will be 25% greater than for a current Claus plant with tail gas clean-up. This assumption is not altogether arbitrary in as much as both processes (electrochemical decomposition technique excepted) require the use of high temperature reactors constructed of hydrogen sulphide resistant materials, condensers, blowers, etc. For the same through-put capacity, the decomposition plant equipment sizings will be smaller since the hydrogen sulphide feed stream is not diluted with combustion air as it is in the Claus; emission control equipment will also be similar in Claus and decomposition plants since many of the potential pollutants are common to all processes.

Based on this assumption, operating costs exclusive of energy (fuel) costs have also been estimated to be 25% higher for the decomposition process. Using these estimates, Table 1 compares the energy requirements, costs, and revenues of the two processes. Rather than including capital and operating costs which vary according to plant size, an incremental cost penalty calculated on a per tonne sulphur produced basis has been assessed against the decomposition process to allow for the estimated increase in these costs. Thus the totals in Table 1 can be considered only on the basis of comparative profitability - common capital and operating costs are not included. It should also be noted that an energy credit (at 80% recovery efficiency) has been given to the Claus process but that this heat, recovered in the form of steam, is not always useable in the processing plant. Therefore this credit, calculated on the basis of full fuel value, is generous to the Claus system. By contrast heat utilization efficiency has been assumed at only 50% for the decomposition reaction. This further discounting against the dissociation system was adopted because of the uncertainties surrounding this process.

TABLE I.

Economic Comparison of the Claus and Decomposition Processes

	Claus	Decomposition
	$3H_2S + 3/2O_2 \rightleftharpoons 3/xS_x + 3H_2O$	$2H_2S \rightleftharpoons 2H_2 + 2/xS_x$
	1 product	2 products
ΔH	-607 to -711 kj	$+159$ kj
ΔH/mole H_2S	~ -218 kj	$\sim +80$ kj
ΔH/tonne S	$\sim -6.85 \times 10^9$ joules	$\sim +2.5 \times 10^9$ joules
Heat input (output) per tonne S	($\sim 5.5 \times 10^9$ joules) Assuming 80% recovery	$\sim 5.0 \times 10^9$ joules Assume 50% reaction efficiency
Value of heat produced (required)/ tonne S (@ $2.50 $\times 10^9$ joules)	$13.75	($12.50)
Product value (assuming S @~$40/ tonne and H_2 @~$.09/m^3)	$40.00	$103.00
Net product value (Heat + product)	$53.75	$ 90.50
Increased capital cost penalty to decomposition process (25% over Claus)	–	($8.00)
Comparative profitability (Common capital and operating costs NOT CONSIDERED)	$53.75	$82.50

However, in spite of this conservative approach, it is clear from Table I that the decomposition process offers considerably more profit potential than the conventional Claus process. Recalling that the totals in Table I represent only the comparative profitability, comparison of the actual profit generated by each process favors the decomposition process more strongly. For example, the total capital and operating costs for a medium sized (150 - 200 tonne/day) Claus plant (including tail gas treatment) are of the order of $32/tonne exclusive of steam credit. Subtracting these costs from the totals in Table I, the overall net revenue from the Claus system would be $21.75/tonne of sulphur produced whereas that from the decomposition process would be $50.50 /tonne - a 230% difference. Even in the unlikely event that the total capital and operating costs for the decomposition process exceed those of the Claus system by 100%, the two processes are still approximately competitive.

As an alternative way of looking at the economics, it is possible to calculate a range of hydrogen production costs for the decomposition process. Using a capital and operating cost of $40/tonne of sulphur produced excluding fuel cost ($32/tonne process cost for Claus processing plus 25% penalty for decomposition) the overall production costs for the decomposition process are about $50.50/tonne of sulphur ($40 + $12.50/tonne fuel cost from Table I). Assuming a by-product sulphur value of $40/tonne, the net cost of the hydrogen produced (~700 m^3/tonne of S) by decomposition is $12.50/tonne of sulphur processed. This corresponds to $12.50/700 m^3 hydrogen giving a net hydrogen production cost of $.018/$m^3$ (~$.50/MSCF). Even if capital and operating costs for the decomposition process reached double the costs of the Claus Process, hydrogen costs from the decomposition system would still be only $.05/$m^3$ (~$1.40/MSCF). From this simple analysis the potential economic benefits of the decomposition process provide considerable incentive for its development.

At present, over 10 million tonnes of sulphur are recovered annually from hydrogen sulphide in North America alone, and this represents a potential source of about 7.5 x $10^9 m^3$ (~250 x 10^9 SCF) of hydrogen. In the future strong growth is foreseen in hydrogen sulphide processing. Conservative estimates suggest that involuntary hydrogen sulphide generation from fuel desulphurization will increase by a factor of 2.5 by the year 2000 in North America, while recent unofficial estimates have claimed that this increase could be ten fold or greater if massive coal gasification and liquifaction projects are brought on stream. There is no doubt that the trend to heavy oils, sour gas, tar sands, shales and coal will require the construction in the next two decades of new hydrogen sulphide processing facilities with

a total capacity well in excess of the current capability.
By 2000, by-product hydrogen sulphide in North America will
represent a potential source of a minimum of 15 x 10^9 m3/year
(5 x 10^{11} SCF/yr) of hydrogen, and potential could be as
great as 25 - 35 x 10^9 m^3/yr (8.5 - 11.5 x 10^{11} SCF/yr).
World-wide the potential is of course even more significant
(Table II).

TABLE II.

Potential Hydrogen Supply from Hydrogen Sulphide (10^9 m^3/yr)

	North America	World (Non-Comm.)
1978	7.5	13.5
1990	12 - 18	20 - 30
2000	15 - 35	30 - 60

With respect to the coincidence of potential sources of
supply and demand, the decomposition concept again seems
attractive. Hydrogen has three major uses including its use
in fuel refining, ammonia production and methanol manufacture.
In general, current and future sources of hydrogen sulphide
are geographically very close to hydrogen markets, and the
potential supply from decomposition could very easily be
absorbed by these markets (in fact, hydrogen from hydrogen
sulphide decomposition could supply about 10 - 20% of demand
in 2000). For refineries and petrochemical complexes, the
process seems ideal. Within the confines of a typical
single plant complex, hydrogen sulphide is produced necessitat-
ing treatment and hydrogen is required for desulphurization
and cracking (in many refineries hydrogen production from
reforming is insufficient and must be supplemented with
product from a hydrogen plant). In contrast to the current
system of fuel desulphurization (Fig. I), the use of the
decompositon process (Fig. II) would conserve resources and
improve profitability.

As we move into an era of higher cost, lower quality
energy, new technologies emphasising efficiency, resource
conservation and energy savings will become increasingly
important. Unconventional processes based on unusual reaction
sequences will become attractive as economic conditions and
energy prices change rapidly. Industries that adjust to
these changes should remain healthy whereas those who do not
may well be left with unprofitable operations.

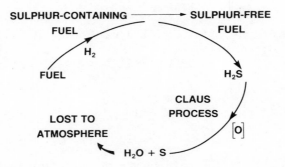

Figure 1.   Fuel desulfurization (present)

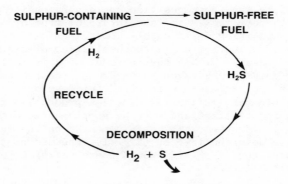

Figure 2.   Fuel desulfurization (novel)

In this light, the hydrogen sulphide decomposition concept deserves careful attention. Both technically and economically it appears to be attractive and its successful development and commercialization will lead to energy savings, resource conservation and improved economic viability for processors of hydrogen sulphide.

Literature Cited

1. Raymont, M.E.D. "The Thermal Decomposition of Hydrogen Sulphide", Ph.D. Thesis, University of Calgary: Calgary, Alberta, 1974.

2. Raymont, M.E.D. "H$_2$ from H$_2$S - A New Use for Hydrogen Sulphide", Proc. Can. Sulphur Sym.: University of Calgary, Calgary, Alberta, 1974.

3. Raymont, M.E.D. Hydrocarbon Proc., 1975, 54 (7), 139.

4. Tanaka, T., Shibayana, R., Kiuchi, H. J. Metals, 1975, 27 (12), 6.

5. Nishizawa, T., Tanaka, Y., Hirota, K. Nippon Kagaku Kaishi, 1977, 7, 929.

6. Kiuchi, H., Tanaka, T. Trans. Soc. Min. Eng. AIME, 1977, 262 (3), 248.

7. Kameyama, T. Ryusan To Kogyo, 1978, 31 (5), 103.

8. Kotera, Y. Int. J. Hydrogen Energy, 1976, 1, 219.

9. Kotera, Y., Todo, T.N., Fukuda, N.K. U.S. Patent No. 3,962,409, June 1976.

10. McBride, R.B., McKinley, D.L. Chem. Eng. Prog., 1965, 61 (3), 81.

11. Dokiya, M., Kameyama, T., Fukuda, K. Denki Kaguku, 1977, 45 (11), 701.

12. Weiner, J.G., Leggett, C.W. U.S. Patent No. 2,979,384, April 1961.

13. Unpublished results.

14. Tanaka, T., Private Communication.

15. Fukuda, K., Dokiya, M., Kameyama, T., Kotera, Y. J. Catal.,
     1977, 49, 379.

16. Fleming, E.D., Fitt, T.C. Ind. Eng. Chem., 1950, 42 (11),
     2249.

17. Bolmer, P.W. U.S. Patent No. 3,409,520, November 1968.

18. Herring, W.O., Jenkins, R. "Standards Support and
     Environmental Impact Statement. An Investigation of
     the Best Systems of Emission Reduction for Sulphur
     Compounds from Crude Oil and Natural Gas Field
     Processing Plants", EPA: Research Triangle Park, N.C.,
     1977.

19. "Standards Support and Environmental Impact Statement.
     Volume 1: Proposed Standards of Performance for Petroleum
     Refinery Sulphur Recovery Plants", EPA: Research Triangle
     Park, N.C. EPA Publication No. EPA-450/2-76-016-a, 1976.

RECEIVED July 12, 1979.

Thermochemical Decomposition of H_2S with Metal Sulfides or Metals

HIROMICHI KIUCHI, TETSUO IWASAKI, ISAO NAKAMURA, and TOKIAKI TANAKA

Department of Metallurgical Engineering, Faculty of Engineering, Hokkaido University Sapporo 060, Japan

Hydrogen plays an important role as a reducing agent in extractive metallurgy. The trend toward energy system improvement based on hydrogen will therefore have great influence on the field of metallurgy and not only with respect to energy source but also to the appearance of new metallurgical processes.

The authors are now studying the simultaneous recovery of metal and sulfur by a combination of the hydrogen reduction of sulfide ore and the decomposition of H_2S to H_2 and S as follows.

$$MS_x + xH_2 = M + xH_2S$$

$$xH_2S = xH_2 + xS$$

$$\overline{MS_x = M + xS}$$

Here the decomposition of H_2S is an extremely important reaction.

Although H_2S is at present a by-product of the desulfurization of fossil fuels on a large scale, only the recovery of free sulfur is carried out by the Claus treatment. On the other hand, the application of H_2S decomposition as an H_2 evolution method is proposed for use in the thermochemical process undertaken for water splitting. Thus, the thermochemical decomposition of H_2S has wide-spread applications in various field.

Methods with metal sulfides

Hydrogen sulfide can fracture into hydrogen and sulfur merely by thermal decomposition, but the equilibrium H_2 concentrations are at best those shown in Table I.

In this study, the experiment based on a combination of two reactions illustrated in Table II was carried out. The equilibrium H_2 concentration generally becomes higher with a decrease in temperature in the sulfurization of the metal sulfides by H_2S, while

0-8412-0522-1/80/47-116-349$05.00
© 1980 American Chemical Society

Table I. The equilibrium H_2 concentrations for H_2S thermal decomposition

Temperature (°C)	400	500	600	700	1000
H_2 concentration (vol-%)	0.1	0.4	1.3	2.8	13.4

Table II. Thermochemical decomposition of H_2S

$$M_xS_y + zH_2S = M_xS_{y+z} + zH_2 \qquad (1)$$

$$M_xS_{y+z} = M_xS_y + zS \qquad (2)$$

$$zH_2S = zH_2 + zS$$

the thermal decomposition of the sulfide proceeds faster at higher temperatures. Therefore, the decomposition efficiency of H_2S will be increased by a combination of reaction (1) at low temperatures and reaction (2) at high temperatures.

The following conditions are necessary for the sulfide used in this cycle. First, reaction (1) has a high equilibrium H_2 concentration at temperatures above 500°C. This is a solid-gas heterogeneous reaction so that the rate may markedly decrease below 500°C. Second, the sulfide formed in reaction (1) should not preferably be a higher sulfide such as polysulfide. A high sulfur activity in the higher sulfide is suitalbe for reaction (2), but a high H_2 concentration can not be expected in reaction (1). Third, reaction (2) can proceed at temperatures below 900°C. This limiting temperature was considered on the assumption of using of an H.T.G.R.(High Temperature Gas Reactor).

Under the above conditions, the use of the non-stoichiometric composition peculiar to sulfides and the use of a monosulfide which can form lower sulfides by the thermal decomposition were considered.

Figure 1 shows the outline of the experimental apparatus used. The sulfide was packed in No.3 in this figure. The H_2 recovery under H_2S flow and the sulfur recovery under argon flow were alternately repeated many times. The thermal decomposition for sulfur was carried out under normal or reduced pressure. In this study, the repeat of the experiment associated with the former was called the normal pressure cycle and that associated with the latter was called the reduced pressure cycle, respectively. Moreover, the H_2 concentration of off-gas was analyzed by gas chromatography and the behavior of the H_2 formation was investigated during the H_2 recovery experiment.

These results in Figure 2 were obtained with iron sulfide used as an example of a non-stoichiometric composition and were obtained repeatedly in both cycles.

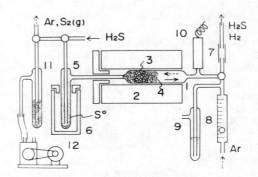

Figure 1. Schematic of experimental apparatus for metal sulfide: (1) reaction tube; (2) electric furnace; (3) metal sulfide; (4) quartz wool; (5) trap; (6) cold bath; (7) sampling tube; (8) flow meter; (9) trap; (10) vacuum detector; (11) trap; (12) vacuum pump.

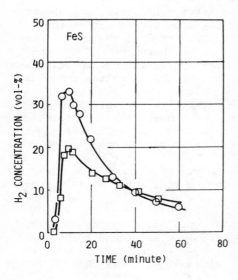

Figure 2. Hydrogen evolution curves for normal and reduced pressure cycles with pyrrhotite (FeS): (□), normal pressure cycle at 600°C; (○), reduced pressure cycle at 550°C.

The influence of the sulfurization temperature was investigated in each cycle. The behavior of each optimum sulfurization temperature is compared in this figure.

The sulfur composition of FeS ranges from FeS_{1+0} to $FeS_{1+0.2}$ at 600°C, and the equilibrium H_2 concentration in that range varies from about 100 % to about 4 %. On the other hand, the composition of the thermally decomposed product was $FeS_{1.11}$ under normal pressure and $FeS_{1.06}$ under reduced pressure. The sulfurized products were both approximately $FeS_{1.2}$. Therefore, the H_2 formation behavior in the figure was explained as showing a concentration corresponding closely to the composition-variation of FeS.

The iron sulfide used in the experiment was obtained from the thermal decomposition of pyrite(FeS_2). The particles were extremely porous, with pores sizes of several tens of microns. In contrast, the reaction behavior obtained with natural pyrrhotite or synthetic FeS composed of fine particles, gave much worse results. Thus, the influence of the specific surface area of a solid on the formation behavior was thought to be important.

In order to understand the characteristics of chalcopyrite, which has a wide range of non-stoichiometric compositions similar to iron sulfide and is a double sulfide, an exeriment with a copper concentrate was carried out.

The concentrate was composed of fine chalcopyrite particles of ca. 50 micron. The maximum H_2 concentration in both the normal and reduced pressure cycles was larger than the value obtained with iron sulfide. The results are shown in Figure 3.

Based on the identification by X-ray diffraction and observation by micrography, the variation was found to be within the non-stoichiometric composition of chalcopyrite in the normal pressure cycle. Despite the decomposition into bornite(Cu_5FeS_4) and pyrrhotite during the reduced pressure cycle, the chalcopyrite was found to be completely restored to its original chalcopyrite form by the succeeding sulfurization.

The sulfur composition of the pyrrhotite was very low, and $FeS_{1.01}$. Since a favorable H_2 formation behavior was not obtained from the experiment using synthetic bornite, the excellent results obtained during the reduced pressure cycle were thought to be due to the pyrrhotite.

Ni_3S_2 is a known lower sulfide as compared to NiS and shows a high equilibrium H_2 concentration over a wide range of sulfur compositions.

The sulfurization of Ni_3S_2 to NiS proceeded easily, though the thermal decomposition of NiS into Ni_3S_2 was found to be difficult under the reduced pressure. The repeated results for the reduced pressure cycle are shown in Figure 4. It shows that consistent behavior is difficult to obtain.

The melting point of Ni_3S_2 is approximately 800°C and that of Ni_3S_{2-x} is 645°C. Accordingly, the melting or sintering of the sulfide in a packed bed occured, and the surface area of the solid was thought to be reduced by each repetition of the thermal decomposition.

The cycle combined with the thermal decomposition at a lower temperature and higher vacuum degree may be suitable for this.

Methods with metals

Generally, the sulfurization of metal exhibits a remarkably high equilibrium H_2 concentration, compared to the sulfurization of metal sulfide as shown in Table III.

Table III. The equilibrium H_2 concentrations for metal sulfurizations

Temp. (°C)	H_2 (vol-%)		
	Bi	Cu	Pb
400	47.3	99.9	99.9
500	27.3	99.9	99.8
600	16.2	99.9	99.6
700	10.2	99.9	99.0

However, few sulfides are capable of being thermally decomposed into metal and sulfur. Noting the decomposibility of Bi_2S_3, Soliman et al([1]) proposed a cycle using Bi. The authors found that Ag_2S decomposed at 800°C, under a reduced pressure of a few mm Hg, to form Ag. The   equilibrium H_2 concentrations for sulfurizations are, however, small for both Bi and Ag. The reaction is hindered by the sulfide film formed on the surface which occurs during the sulfurization of solid metal.

In this study, the use of liquid metal was examined. A smelting reaction was considered for the recovery of metal from the sulfide. Since sulfur changed to SO_2 in this case, the reaction of SO_2 and H_2S by the Claus reaction was assumed.

The reaction equations of the cycle using liquid Pb are shown in Table IV. As a means of preparing Pb from PbS, the roast-reaction or air-reduction method is well-known in nonferrous extractive metallurgy. The reactions for this method can be presented as follows :

$$2PbS + 3O_2 = 2PbO + 2SO_2$$

$$2PbO + PbS = 3Pb + SO_2$$

The authors have already found the direct production of Pb from PbS by oxidation under low oxygen partial pressure.([2]) Accordingly, the experiment of H_2 formation with lead was carried out.

Table IV. Two methods with Pb(l)

$$Pb(l) + H_2S = PbS + H_2 \qquad\qquad (3)$$

$$PbS + O_2 = Pb(l) + SO_2 \qquad\qquad (4)$$

$$SO_2 + 2H_2S = 2H_2O + 3S \qquad\qquad (5)$$

$$(3)+(4): \quad H_2S + O_2 = H_2 + SO_2$$

$$(3)+(4)+(5): \quad 3H_2S + O_2 = H_2 + 2H_2O + 2S$$

Figure 5 shows an outline of the experimental appratus. The fused lead was placed in a reaction tube made of quartz. The reaction was studied by the bubbling method in which H_2S was bubbled into the fused lead and by a soft blowing method in which H_2S was blown onto the surface of the lead. The H_2S gas stored in vessel No. 1 was circulated by pump No. 2. Since the lead content was in excess of the H_2S in order to maintain the liquid state, the reaction behavior was examined in this experiment until the gas reached the equilibrium composition.

The equilibrium H_2 concentration was 99.8 % at 500°C and 97.9 % at 800°C. The value is very high, in spite of the temperature being high. As a result, the higher the reaction temperature, the more favorable the H2 formation behavior in this reaction system. In addition, the sulfurization of Pb is an exothermic reaction and a slight rise in temperature was observed during the experiment. However, the acceleration of the reaction at low temperature was also examined.

Figure 6 shows the result obtained by the bubbling method. Since the conversion of the ordinate are in proportion of formed H_2 by the reaction, it should reach the same value as the equilibrium H_2 concentration. The value is approximately 99 % at this temperature.

Very small amounts of various metals were added to the lead to accelerate the reaction rate. As a result, Ni was found to be an effective metal. The critical amount of the effective Ni was about 1 wt-%, which corresponds to the solubility of Ni in fused Pb. Therefore, the acceleration of the reaction was obviously dependant to the Ni dissolved into the lead.

The effect was also confirmed as being maintained after the completion of Ni sulfurization, even if a simultaneous sulfurization of Ni was assumed. Consequently, it was thought that the dissolved Ni did not merely participate in the prior sulfurization but acted catalytically.

The results obtained by a gas sweeping method were rather poor compared to those obtained by the bubbling method. In particular, the reaction almost stagnated after a time of about 30

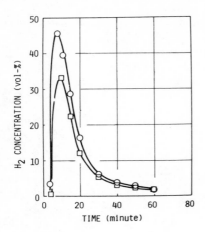

Figure 3. *Hydrogen evolution curves for both cycles with Cu concentrate composed of chalcopyrite (CuFeS₂): (□), Cu concentrate; (○), reduced pressure cycle.*

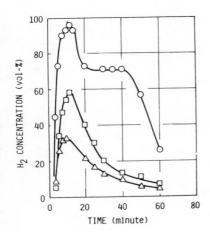

Figure 4. *Hydrogen evolution curves for Ni₃S₂ at 550°C at the reduced pressure cycle: (○), 1st; (□), 2nd; (△), 6th.*

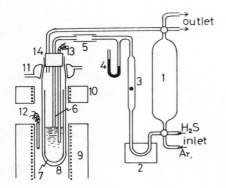

Figure 5. *Schematic of experimental apparatus for molten lead: (1) gas chamber; (2) circulation pump; (3) flow meter; (4) pressure gauge; (5) sampling tube; (6) blowing nozzle; (7) quartz tube; (8) reaction tube; (9) electric furnace; (10) electric furnace; (11) water jacket; (12) thermocouple; (13) thermocouple; (14) silicon stopper.*

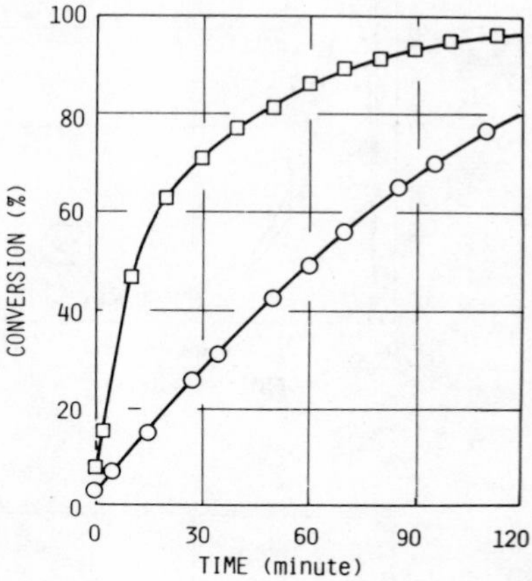

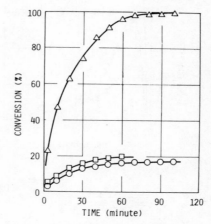

Figure 6. Effect of Ni addition to molten lead on the conversion of H_2S to H_2 by the bubbling method at 550°C: (○), Pb; (□) 1 wt % Ni–Pb.

Figure 7. Effect of Cu addition on the conversion of H_2S to H_2 by the soft-blowing method at 600°C: (△), Pb–Cu; (□), Pb–Ni; (○), Pb.

minutes had elapsed.   From observation of the liquid surface
during the reaction, the surface at that time was found to be covered with a film of the lead sulfide formed. The formation of this
film was also seen in lead containing added Ni, and no accelerating
effect could be found after this time with agitation.

The addition of Cu, however, resulted in an outstanding effect as shown in Figure 7. The amount of Cu addition was limited
to that corresponding to the solubility of about 1 wt-%. In this
case, too, the behavior was considered to involve catalytic action,
instead of the prior or simultaneous sulfurization.

In addition, the surface observation showed that the sulfide
formed with Cu addition did not cover the surface but accumulated
by swelling from the surface. The sulfide accumulated was extremly
porous. Consequently, the dissolved Cu was considered to have effect on the growth of PbS crystals.

Summary

The decomposition efficiency of H_2S was increased by a reduced pressure cycle of a few mm Hg for metal sulfides.

The utilization of a metal may be more promising as a H_2
recovery method rather than the decomposition of H_2S.

As a primary approach toward "Hydrogen Economy", the H_2S
by-product obtained from fossil fuels or by extractive metallurgy
of sulfide ores should be considered as more important for H_2
recovery than for the sulfur which can be recovered.

References

1. Soliman,M.A., Calty,R.H., Conger,W.L., Funk,J.E., Can.J.Chem.
Eng., 1975, 53, 164–169.
2. Kiuchi,H., Tanaka.T., Trans.Soc.Min.Eng.AIME, 1977, 262, 248–
254.

RECEIVED July 12, 1979.

The Sulfur-Cycle Hydrogen Production Process

G. H. FARBMAN and G. H. PARKER

Westinghouse Electric Corporation, Advanced Energy Systems Division,
P.O. Box 10864, Pittsburgh, PA 15236

Changes in sources of energy, technology, and economic conditions can have substantial effects upon the way we meet industry's
need for hydrogen. Such changes may require the substitution of new
techniques in place of today's major reliance on fossil fuels for
the production of hydrogen (e.g., steam methane reforming, partial
oxidation of liquid fuels). Future fossil fuel costs, availability,
and environmental effects – be they real, perceived, or political
in nature – may dictate that hydrogen, whose needs are projected
to grow substantially during the coming decades, be produced from
water with solar or nuclear energy sources. Since one cannot predict with any certainty when such a change may be required, and re-
cognizing the long lead-time necessary to bring a new technology
"on-line," it is prudent that research, development, and demonstration activities in new, advanced, non-fossil energized hydrogen
production techniques be undertaken now. It is on this basis that
Westinghouse, and others in the United States and elsewhere, are
participating in programs in advanced hydrogen production.

The water splitting technique which Westinghouse has under
development is a closed cycle hybrid electrochemical/thermochemical
system called the Sulfur Cycle Hydrogen Production Process. The
process is capable of operating with high temperature nuclear or
solar heat sources and is expected to be able to produce hydrogen
at an overall thermal efficiency, including the inefficiencies
associated with the generation of the required electric power, of
close to 45 percent.

In the Sulfur Cycle, hydrogen is produced in a low temperature electrochemical step, wherein sulfuric acid and hydrogen are
produced from sulfurous acid, i.e.,

$$2 H_2O + SO_2 \rightarrow H_2SO_4 + H_2 \qquad [1]$$

The cycle is closed by the high temperature dissociation of sulfuric acid to sulfur trioxide and water, and the subsequent reduction of sulfur trioxide to sulfur dioxide and oxygen, i.e.,

0-8412-0522-1/80/47-116-359$07.75
© 1980 American Chemical Society

$$H_2SO_4 \rightarrow H_2O + SO_3 \rightarrow H_2O + SO_2 + \tfrac{1}{2} O_2 \qquad [2]$$

The net result of Reactions 1 and 2 is the decomposition of water into hydrogen and oxygen. Sulfur oxides are involved as recycling intermediates. Although electrical power is required in the electrochemical step, much smaller quantities then those necessary for water electrolysis are needed. The theoretical voltage to decompose water is 1.23 V, with many commercial electrolyzers requiring over 2.0 V. The power requirements for Reaction 1 (0.17 volts at unit activity for reactants and products) are thus seen to be theoretically less than 15 percent of those required in conventional electrolysis. The cell voltage of 600-800 mV believed to be obtainable in a practical sulfurous acid electrolyzer would only require some 30-40 percent of the electric power of advanced water electrolyzers. This dramatic change in the heat and work required to decompose water, compared to water electrolysis, can lead to a technically and economically attractive non-fossil fuel energized hydrogen production system.

The process is shown schematically in Figure 1. Hydrogen is generated electrolytically in an electrolysis cell which anodically oxidizes sulfurous acid to sulfuric acid while simultaneously generating hydrogen at the cathode. Sulfuric acid formed in the electrolyzer is then vaporized, using thermal energy from a high-temperature heat source. The vaporized sulfuric acid (sulfur trioxide-steam mixture) flows to an indirectly heated reduction reactor where sulfur dioxide and oxygen are formed. Wet sulfur dioxide and oxygen flow to the separation system, where oxygen is produced as a process co-product and the sulfur dioxide is recycled to the electrolyzer.

Process parameters for a commercial system will evolve as the technology matures. Conceptual design studies (References 1,2,3) and experimental analyses have, however, led to the selection of ranges of interest and objectives for major process parameters that should result in a technically sound and economically competitive system. Table I summarizes these values.

This paper will discuss the development status of the Sulfur Cycle Hydrogen Production Process, the results of evaluations of process performance, and the program steps which can lead to commercialization of the system.

Development Status

The program for the development of the Sulfur Cycle has been underway since 1973, with Westinghouse funding, and has been supported by the Department of Energy since 1976. The development program plan, as currently conceived, is a multiyear program leading to a pre-pilot integrated bench scale cyclic process development unit (PDU) by 1983, followed by a pilot scale and/or demonstration units at a later date.

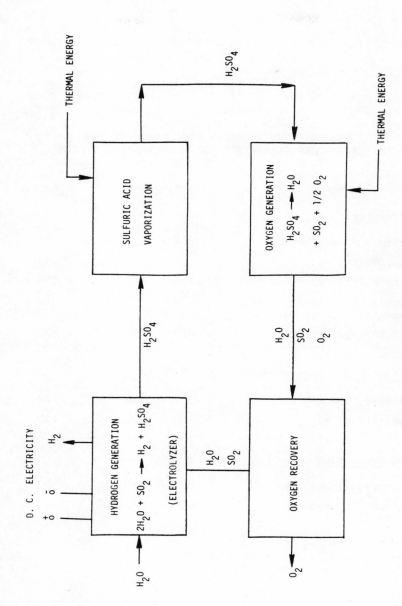

Figure 1. Hydrogen generation schematic design

TABLE I.

SULFUR CYCLE HYDROGEN PRODUCTION SYSTEM
MAJOR PROCESS PARAMETERS FOR A COMMERCIAL SYSTEM

Parameters	Range	Objective
Electrolyzers		
Pressure, Atmospheres	3.5 – 30	20
Temperature, ^{o}C	50 – 125	100
Current Density, mA/cm^2	100 – 300	200
Electrolyte Sulfuric Acid Concentration, wt Percent	50 – 70	55
Cell Voltage, mV	450 – 1000	600
Sulfuric Acid Vaporization		
Pressure, Atmospheres	3.5 – 30	20
Temperature, ^{o}C	360 – 470	450
Oxygen Generation		
Pressure, Atmospheres	3.5 – 30	20
Maximum Temperature, ^{o}C	760 – 925	870

Table II indicates the four major areas to be treated in the overall development program. These are the Electrolyzer, Sulfur Trioxide Reduction, Integrated Testing, and Process Studies. The sections below describe the current status and the effort to be accomplished in each area for the duration of the program.

Electrolyzer. The SO_2 depolarized electrolyzer is the component in which hydrogen is produced. The development program is organized to consider the electrocatalysts, electrodes, cell separators, pressurized and bipolar cell design and performance, and scale-up of test articles to demonstrate the potential of meeting performance and economic objectives. The most significant progress made to date has been in the areas of electrocatalysts, single cell performance, and bipolar cell design and performance.

The electrochemical reaction proceeds most effectively in the presence of a catalyst, and the nature of the catalyst can have a significant effect upon the electrode overpotentials. As a matter of convenience, all of the early work in the electrolyzer development used platinum as both the anode (SO_2 oxidation electrode) and cathode (H_2 generation electrode) catalyst. It was recognized, however, that although platinum might be a technically satisfactory catalyst for the cathode, it was only marginally suitable as the anodic catalyst.

A program of catalyst identification and evaluation was therefore undertaken. The first step in assessing electrolytic activities for the SO_2 oxidation reaction was to measure open circuit potentials of several candidate materials in a SO_2-saturated sulfuric acid solution at room temperature. It is generally accepted that good electrocatalysts have low open-circuit potentials. Figure 2 shows the results obtained as compared to reference platinum and platinum-black materials. As can be seen, an electrode made of WAE-3 exhibits a markedly reduced potential compared to reference platinum and gold materials. A Tafel plot for this material (Figure 3) shows that the limiting current density for a solid electrode made of WAE-3 is substantially greater than that for a platinum electrode. Since the effective surface area of the porous carbon electrodes which would be used in a cell, with WAE-3 distributed as a catalyst, would be at least 100 times greater than the effective area of the solid electrode of Figure 3, it is clear that the objective of current density for the commercial cell can be met.

A parameter of significant importance in cell performance is the operating temperature. Figure 4 shows the measured effect of temperature on the electrode potential for the WAE-3 electrode of Figure 3. As can be seen, a distinct decrease in anode overpotential occurs as temperature is increased from 25 to 90°C. It should be noted that the pressure of the cell was kept constant, at atmosphere pressure, over the range of temperatures. This constant pressure, while temperature is increasing, results in a substantial drop (from ~1M to 0.16M) in the solubility of sulfur dioxide in the electrolyte. The decreased sulfur dioxide content precludes

TABLE II.

SULFUR CYCLE HYDROGEN PRODUCTION PROCESS
DEVELOPMENT AREAS

- Electrolyzer (Hydrogen Generation)

    Catalysts

    Electrode Materials and Fabrication

    Separator Materials

    Pressurized Cell Performance

    Electrolyzer Design Optimization

    Scale-Up Tests

- Sulfur Trioxide Reduction (Oxygen Generation)

    Catalysts

    Pressurized Performance

    Component Design Optimization

    High Temperature Materials

    Scale-Up Tests

- Integrated Testing

    Laboratory Model

    Process Development Unit (PDU)

    Pilot Plant

- Engineering Studies

    Process Flow Sheet Evaluation

    Heat Source Interface Studies

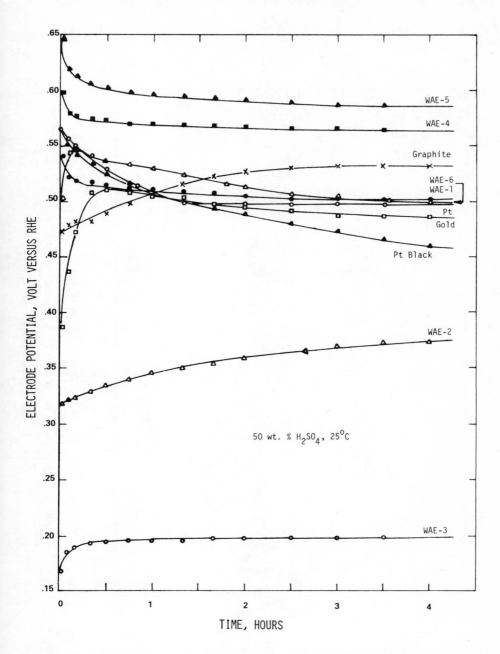

Figure 2.   The variation of open circuit potentials on various electrodes as a function of time (SO₂ was bubbled into the electrolyte)

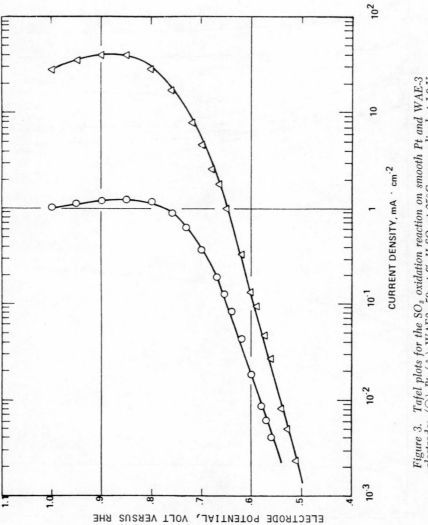

Figure 3.   Tafel plots for the SO_2 oxidation reaction on smooth Pt and WAE-3 electrodes: (○), Pt; (△), WAE3. 50 wt % H_2SO_4 at 25°C, preanodized at 1.0 V for 30 min. Magnetic stirring.

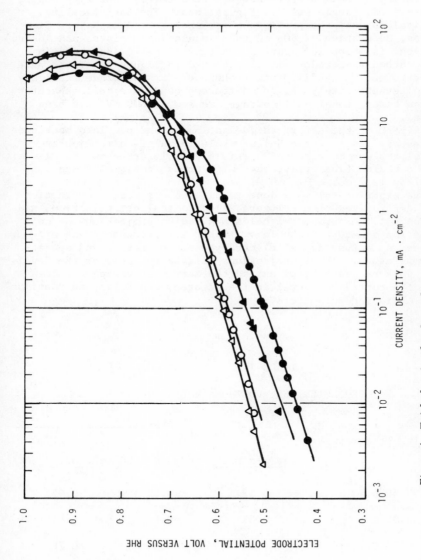

Figure 4. Tafel plots for the electrochemical oxidation of SO_2 on preanodized electrodes in 50 w/o H_2SO_4 solutions at temperatures of: 25°C (△); 50°C (○); 70°C (▲); and 90°C (●). Atmospheric pressure, SO_2 saturated, and magnetic stirring.

any improvement in limiting current density - an increase in
pressure will increase the sulfur dioxide solubility and provide
a greater current density capability.

Testing of cell performance has been carried out with small,
unpressurized, single cell configurations. The test facility,
which includes all the provisions for anolyte and catholyte circulation, saturation of SO_2 in the circulating anolyte, and instrumentation, is shown in Figure 5. A typical cell, incorporating
porous carbon electrodes, is illustrated in Figure 6. Tests have
been performed using different techniques for electrode fabrication, alternate catalysts, and different candidate cell separators.
In these tests, total cell voltages as low as 600 mV have been
measured.

As part of the effort to design, build and put into operation
a laboratory model of the Sulfur Cycle, a multi-cell, bipolar
electrolyzer was constructed. This electrolyzer, shown in Figure 7, contains five cells, each with cross-sectional dimensions
of 12.7 by 12.7 cm.

The experimental work done to date in the reaction kinetics
of SO_2 depolarized electrolyzers, electrocatalyst evaluation, and
tests on small experimental cells has led to projections of cell
voltage for a mature technology. Table III presents these projections, as a function of electrolyte concentration and operating temperature. These projections will be refined, as the development work progresses, to expand the matrix of variables, increase the confidence level in the figures, and define an "optimum"
set of operating parameters.

TABLE III.

PROJECTED CELL VOLTAGE (VOLTS)*

Operating Temperature	Electrolyte		
	50 w/o H_2SO_4	60 w/o H_2SO_4	70 w/o H_2SO_4
50°C	0.57	0.69	0.97
90°C	0.53	0.62	0.84
125°C	0.51	0.59	0.78

* Pressure - 20 atm

Current Density - 100 mA/cm^2

Figure 5. Electrolyzer test cell facility

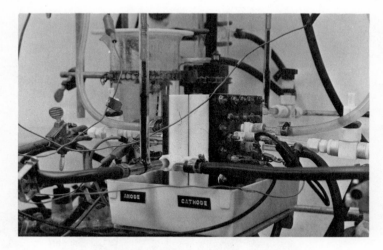

Figure 6.   Small electrolyzer test cell

*Figure 7.   Laboratory model 5-cell bi-
polar electrolyzer*

Sulfur Trioxide Reduction. The effort on Sulfur Trioxide
Reduction covers that part of the hydrogen generation process
that takes sulfuric acid from the electrolyzer, concentrates and
vaporizes the stream, and reduces the resulting steam/SO_3 mixture
to steam, SO_2, O_2, and unreacted SO_3. The scope of this task includes not only the catalyst investigations for the reduction
reaction, but also the materials of construction for the high
temperature acid vaporizer and reduction reactors. The applicability of this part of the process to other end uses, such as the
use of the endothermic reduction reaction as a means to store
energy, requires that sufficient flexibility be designed into the
experimental program to assure data applicable to a spectrum of
applications.

The continuing capability to quickly and inexpensively screen
candidate reduction catalysts to be able to determine those with
high potential for economically attractive activities and life-
times is an important aspect of demonstrating the merit of the
process. Accordingly, atmospheric pressure test facilities, such
as shown in Figure 8, are employed to assess the activities and
aging characteristics of selected catalysts over the temperature
range of interest. There are two such facilities currently in
use. One is used to determine the activity of selected catalysts
and catalyst shapes over a range of temperatures and space velocities. The other is available to evaluate aging characteristics
of catalysts in nominally 1000 hour steady state tests. The inlet
process fluid in all cases contains SO_3 and water vapor in the
proportion that simulates the inlet conditions for a desired process sulfuric acid concentration. The tests can also be performed
with anhydrous SO_3 to simulate the reactions of a sulfur trioxide
energy storage and transport system.

Under this program, testing of several catalysts has been
undertaken, with platinum so far showing the highest potential
performance worth. Figure 9 shows typical data for a platinum
catalyst at several space velocities over a temperature range of
$650^\circ C$ to $900^\circ C$. It is interesting to note that the conversion of
SO_3 to SO_2 and O_2 is nearer to equilibrium values at the higher
space velocity. This is attributed to the effects of more turbu-
lence in the flow stream and enhanced mass transfer of species to
and from the catalyst.

While the current catalyst screening is done at atmospheric
pressure, commercial reactors would operate at pressures up to
perhaps 20 atmospheres. The effect of pressure on the decomposition kinetics has to be explored. Therefore, a pressurized catalytic sulfur trioxide reduction reactor has been designed and
plans for its construction and operation have been made.

As in most advanced technologies, materials of construction
are challenged by the Sulfur Cycle process. The most critical
structural and heat transfer materials problem is in the containment of boiling sulfuric acid, at elevated pressures and temperatures, during the vaporization prior to SO_3 reduction. The

Figure 8.   SO₃ reduction test facility

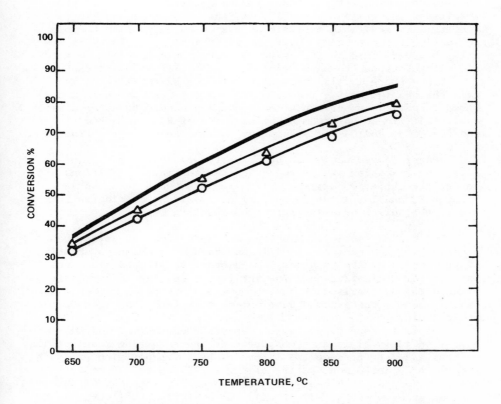

Figure 9.   Extent of reaction for space velocities of 1000 hr^{-1} ($\bigcirc$) and 5000 hr^{-1} ($\triangle$) for catalyst MB-3 (1% Pt on a spherical alumina substrate). $SO_3 \rightarrow SO_2 + \frac{1}{2}O_2$. (——), Equilibrium (calculated). Atmospheric pressure, inlet stream. Volume fraction: SO_3, 0.11; steam, 0.26; carrier gas (argon), 0.63.

operating conditions are well beyond those normally used in the manufacture or use of sulfuric acid and, therefore, materials data are generally unavailable. Work is underway to screen various candidate materials to determine their suitability or need for new alloy development.

The screening is being done in 98 w/o sulfuric acid. Materials samples are placed in the liquid and vapor phases of the test vessel, as well as at the liquid/vapor interface. Typical test results are shown in Figures 10, 11, and 12. As is apparent, commercially available metallic alloys do not appear to be able to survive the operating environment. Silicon containing materials, however, used as structures or protective coatings on metallic alloys, appear to have promise of fulfilling the process needs.

Since there is also a concern with the suitability of materials for use in the SO_3 reduction reactor, a task has been undertaken to provide experimental screening of candidate materials for application in an environment of mixtures of SO_3 and steam (inlet) and SO_3, SO_2, and O_2 and steam (outlet) and temperatures from $425^{\circ}C$ to $900^{\circ}C$.

Materials evaluation is being performed in two test facilities, one operating at the reduction reactor inlet conditions and one at the outlet conditions. Test results, such as shown in Figures 13 and 14, identify several materials as candidates for use in the sulfur trioxide reduction reactor.

Integrated Testing. Although the individual steps of the process have been demonstrated in laboratory experiments, testing of the entire system is necessary in order to proceed along the path to commercialization. The integrated testing is currently planned to be comprised of three sequential steps, i.e., the laboratory model, the process development unit (PDU), and the "pilot-scale" unit.

The laboratory model, which was designed, built, and put into operation in 1978, is an atmospheric pressure working model of the process sized to produce two liters of hydrogen per minute. Such a model provides demonstration of the cyclic operation of the process, provides the capability to assess interactions of the various process steps, and provides an operating facility that can support the technical feasibility of hybrid electrochemical-thermochemical hydrogen production processes. The laboratory model, shown schematically in Figure 15, includes all the steps of the process, but does not make any attempt to demonstrate overall process efficiency nor materials of construction for a "commercial" unit. The model uses electrical heaters wherever thermal energy is required and all waste heat is discharged with no attempt at recovery. The material of construction for the high temperature parts of the model is primarily quartz. A photo of the model is shown in Figure 16, while the control and data acquisition system is shown in Figure 17. The model has been successfully operated and is currently used to test, on a larger scale than in laboratory

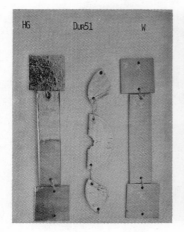

Figure 10.   Materials samples after exposure to boiling 97.7 w/o sulfuric acid at 411°C: left, after 250 hr; right, after 500 hr. HG = HastelloyG; DUR51 = Durichlor51; and W = tungsten.

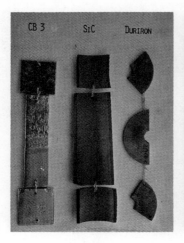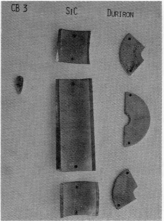

Figure 11.   Materials samples after exposure to boiling 97.7 w/o sulfuric acid at 411°C: left, after 250 hr; right, after 500 hr.   CB 3 = Cartech Type 20CB 3; SiC = chemical vapor deposited silcon carbide; Duriron = Duriron.

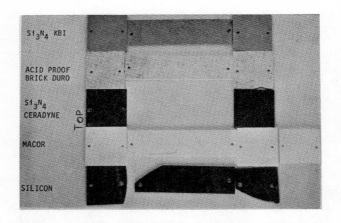

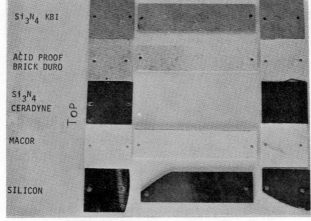

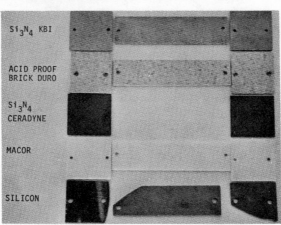

Figure 12. Materials samples after exposure to boiling 97.7 w/o sulfuric acid at 411°C: left, 0 hr; middle, 250 hr; and right, 500 hr.

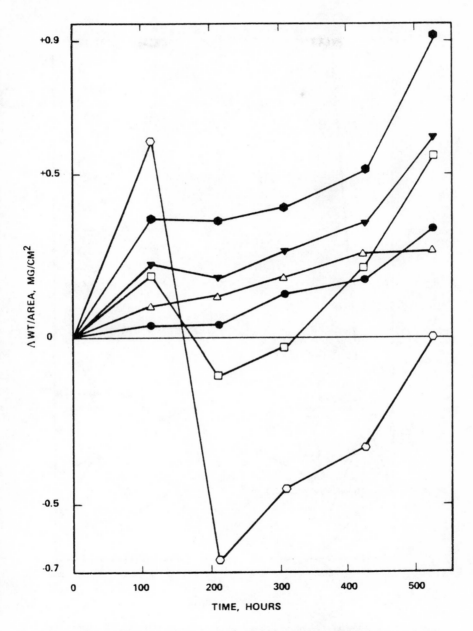

Figure 13. Corrosion behavior of reduction reactor materials samples in anhydrous environment: (○), Hastelloy C276; Cartech CB3; (▼), Incoloy 825; (△), Inconel 625; (◇), SS 310; and (⬣), SS 18-18-2. Furnace temperature, 482°C (900°F); anhydrous SO$_3$ 12 cc/min; argon 128 cc/min. Erratic erosion rate behavior of SS 310 and Cartech 20CB3 is caused by the spalling of the corrosion product. Negative values indicate weight gain per unit area.

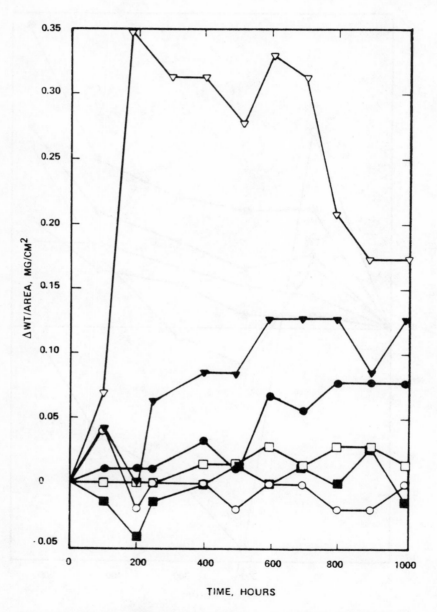

Figure 14. Corrosion behavior of reduction reactor materials samples: ($\triangledown$), Inconel 625; ($\bigcirc$), silicon; ($\square$), silicon nitride; ($\bullet$), alonized Inconel 62; ($\blacksquare$), silicon carbide; and ($\blacktriangledown$), Inconel 657. Furnace temperature, 482°C; SO_3, 25 scc/min; steam, 58 scc/min; argon, 78 scc/min.

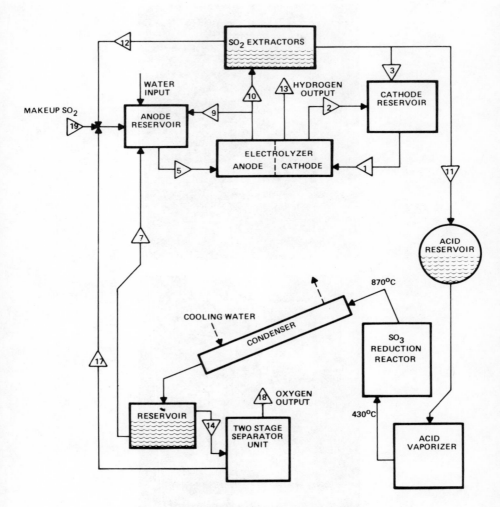

Figure 15.   *Laboratory model schematic*

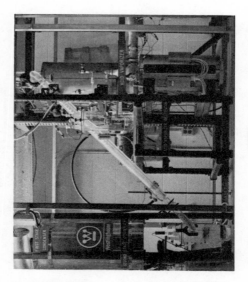

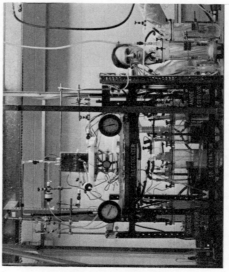

Figure 16.   Photo of laboratory model

hoods, sulfur trioxide reduction catalysts and electrolyzer mechanical configurations.

The purpose of the PDU task may be simply stated - to demonstrate, with an integrated bench scale unit, closed-cycle operation of the process at operating conditions close to that desired
for a commercial unit. The demonstration, as currently conceived,
will include a multicell pressurized SO_2 depolarized electrolyzer
which converts sulfurous acid and water into hydrogen and sulfuric
acid; an acid vaporizer for converting aqueous sulfuric acid into
steam and sulfur trioxide and a thermal reduction reaction for
catalytically reducing sulfur trioxide into sulfur dioxide and
oxygen; and a recovery system for recycling water, sulfur dioxide,
and unreacted sulfuric acid to the electrolyzer. This equipment
will be capable of operating continuously for hundreds of hours
in a fully closed cycle at pressure and temperatures typical of
those expected on commercial units.

The equipment design concepts, catalysts, and materials
employed in the PDU will, to the greatest extent possible, be the
same as those which would be used in commercial systems. The
electrolyzer will use those anodes, cathodes, membranes and mechanical arrangements deemed most appropriate from the earlier testing. The most promising reduction catalyst will be used for con-
ducting the sulfur trioxide reduction reaction. The materials of
construction of both the acid vaporizer and reduction reactor will
be selected based upon the results of the substantial materials
evaluation programs. It is judged that the integrated bench scale
PDU will confirm not only the capability of process operation in
a cyclic fashion, but also the efficiency levels and engineering
approaches of a larger scale pilot or commercial unit.

The pilot scale unit is expected to demonstrate the overall
operability and performance of the process under conditions
approaching that which may occur in commercial units. Neither
the size nor the schedule for the pilot demonstration has yet been
established.

Process Studies. The value of a new technology and the emphasis on development activities are very much related to the assessment of the place that technology can fill in a commercial environment. Accordingly, effort has been expended throughout the
Sulfur Cycle program to develop flow sheets, mass and energy balances, and economic projections for the process to determine if
the Sulfur Cycle, when developed and operating at conditions considered reasonable, can be a commercially acceptable system for
the production of hydrogen. The results of these analyses also
provide valuable information on the importance of certain operating
parameters on the overall systems performance and, therefore, the
relative emphasis that should be placed on those parameters in the
development activities.

A number of process evaluations have been performed, in varying amounts of detail, to reflect the use of the process with very

high temperature nuclear heat sources, high temperature solar heat
sources, and low temperature heat sources. The kind of heat source
determines both the size of hydrogen production plant that one can
reasonably contemplate and the timing for commercial introduction.
For example, the use of the process for large scale merchant hydrogen production will require a very high temperature nuclear reactor
(VHTR) as an energy source. The development activities on the
VHTR, both in the United States and abroad, indicate the commercialization of this nuclear-hydrogen plant would be post-2000. If
a solar thermal energy source is considered, commercialization
perhaps can be attained during the 1990's, but the size of production plant is limited to what may be an economically practical
heliostat field. For near-term implementation, in small units, a
modified version of the Sulfur Cycle, using currently available
energy sources, can be commercialized by the mid to late 1980's.

The nuclear-energized process evaluations have progressed
from the first conceptual design (1) through an updating that considered both nuclear and non-nuclear heat sources (2) to the most
recent updating (3). The very high temperature nuclear reactor
(VHTR) provides all the thermal and electrical energy requirements
for the entire plant. The conceptual designs were based upon the
production of 380 million SCFD of hydrogen at a pressure of 20
atmospheres. The peak temperature in the sulfur trioxide reduction (oxygen production) step is $871^{o}C$.

The overall process efficiency, defined as the higher heating
value of the hydrogen produced divided by the total thermal energy
input to the plant, is calculated to be 46.8 percent when operating
at an electrolyzer cell voltage of 600 mV. This cell voltage is
believed to be a reasonable objective for the Sulfur Cycle program.
Figure 18 shows the variation to be expected in overall efficiency
as a function of cell voltage.

The economics of the Sulfur Cycle plant energized by the VHTR
have been evaluated, using a 1976 cost basis, in a manner similar
to that of earlier assessments (1,2). The hydrogen production
cost is made up of the contributions of nondepreciating and depreciating capital, operation and maintenance, and nuclear fuel cycle
costs. These are calculated on an annual basis. The annual charge
on nondepreciating assets; e.g., land, is assumed to be 10 percent.
The charge on depreciating assets is 18 percent. An 80 percent
capacity factor is assumed. No credit is taken for the oxygen
produced. Figure 19 shows the variation of the resultant hydrogen
"gate price" with cell voltage and nuclear fuel costs. It is interesting to note that by doubling the nuclear fuel cost in
Figure 19, the hydrogen "gate price" increases only about 13 percent.

Further work in process evaluation has been done to evaluate
the potential performance of the process when energized by a solar
heat source. These studies, which were of a preliminary nature,
were based on the solar hydrogen plant being integrated into a
chemical process to produce a currently marketable commodity.

Figure 17.   Photo of laboratory model control station

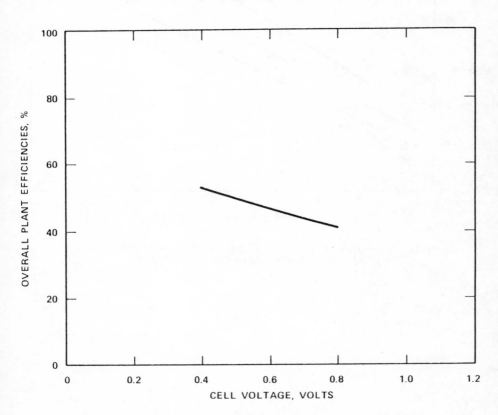

Figure 18.   Sulfur-cycle overall plant efficiency vs. cell voltage

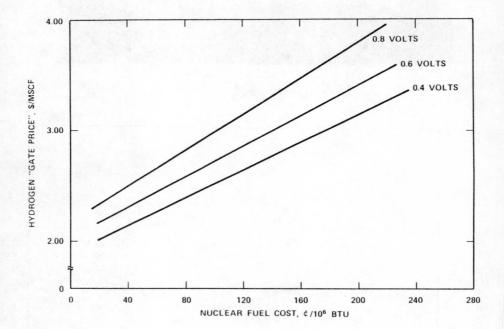

Figure 19. Sulfur-cycle plant hydrogen gate price vs. nuclear fuel cost—1976 basis with cell-operating voltage as a parameter

Chemical processes investigated included the production of ammonia, methanol, hydrogen peroxide, and the reduction of iron ore.

As an example, Figure 20 schematically illustrates the concept of a solar-ammonia plant, with the Sulfur Cycle as the source of hydrogen for the ammonia synthesis. The solar portion of the plant includes sufficient energy storage facilities to provide electrical power for the 24 hour per day operation of the hydrogen and ammonia systems. The high temperature portions of the Sulfur Cycle are oversized, with storage provided for the reduction reaction products, to permit sufficient processing during the hours of insolation to meet the around-the-clock operation of the balance of the system. The projected economics of this plant, which is significantly smaller than the conventional ammonia plants being built today, indicate ammonia production costs that are several times that of the larger conventional natural gas energized production plants at today's natural gas price.

Finally, preliminary scoping studies have been done on the near-term application of the Sulfur Cycle. For the near term, it is assumed that there is no very high temperature nuclear or solar heat source, and the process is therefore modified to operate without the need for high temperature heat. This modification, dubbed the "open-cycle," employs only the electrolytic step of the cycle. Feed materials to the electrolyzers are water and sulfur dioxide, with the latter coming from such sources as flue gas clean-up systems, smelters, or oxidation of sulfur. In the electrolyzer, sulfuric acid, at a concentration of perhaps 50 w/o, and hydrogen are produced. The electrical energy required can come from any power plant, i.e., fossil, nuclear, solar, etc. Relatively low temperature thermal energy is used to concentrate the sulfuric acid to the degree desired for economic shipping. The process, therefore, becomes one for producing sulfuric acid and hydrogen – two products of commercial value – and can be commercialized as soon as the low voltage sulfurous acid electrolyzer development program is completed.

In the scoping studies, a large smelter was assumed to be the source of sulfur dioxide. As shown in Figure 21, the feed of 1000 tons per day of SO_2 into the plant would, neglecting losses, result in the production of 1530 tons per day of fully concentrated sulfuric acid and 11.8 million SCF per day of hydrogen. A preliminary evaluation of the costs of production was made assuming that the SO_2, as a waste product from the smelter, was provided at no cost other than the incremental costs, over that required for its use in a contact sulfuric acid plant, to clean it to the extent required by the electrolyzer. Figure 22 depicts both the cost of the hydrogen as determined by the value assigned to the sulfuric acid produced, and the cost of sulfuric acid as determined by the value assigned to the hydrogen. Although the design and costing were not done in detail, it is apparent that the approach has merit as both a hydrogen and sulfuric acid production plant.

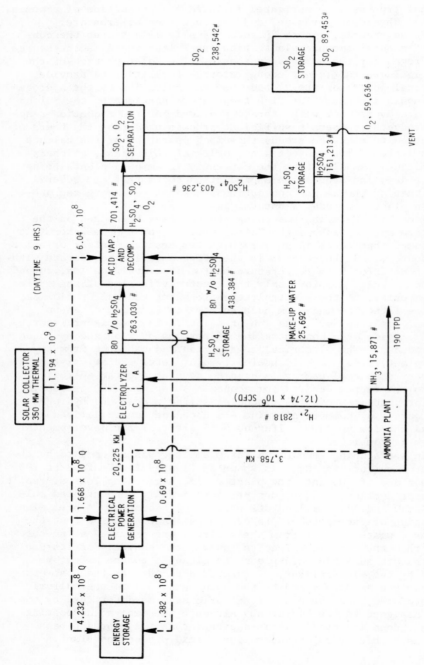

Figure 20. Schematic of solar ammonia plant: # = flow rate in pounds/hr; Q = heat content in Btu/hour.

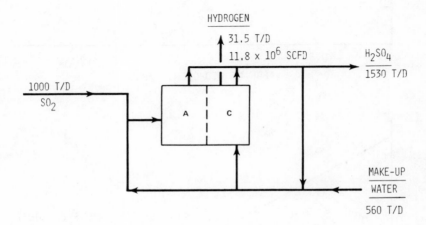

Figure 21.   Schematic of open- cycle sulfur cycle

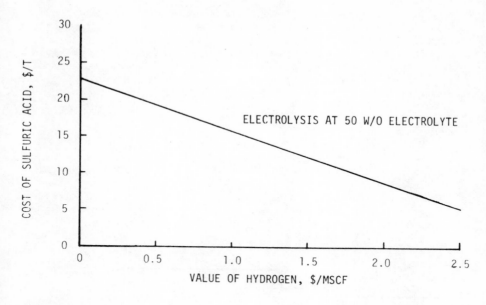

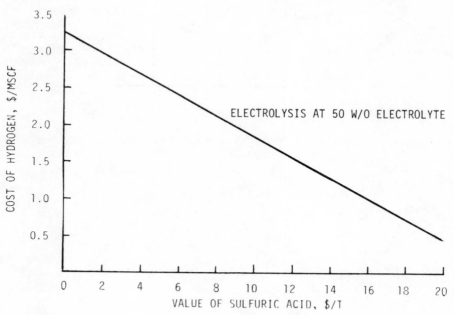

Figure 22.   Open-cycle economics

Similar "open-cycle" scoping studies using waste SO_2 from combustion flue gases and SO_2 produced from sulfur show equivalent economic potential. The latter, of course, would have a higher sulfuric acid production cost because the sulfur feedstock would have an acquisition cost that must be included in the economics. Nevertheless, the cost of sulfuric acid, when tempered by the revenue obtained from the sale of hydrogen, can be extremely attractive. The reverse is also true.

Conclusions

The performance characteristics of the Sulfur Cycle Hydrogen Production Process, as identified in the process evaluations, indicate that the system is capable of effectively filling a need for efficient hydrogen production. The experimental and analytical work being performed on the Sulfur Cycle continues to support the technical and economic potential of the system as an efficient and cost effective way to produce hydrogen. The development program, as currently formulated, is predicated on developing and demonstrating the technology on a timely basis to meet the needs of a phased commercialization plan. A scenario is foreseen where sulfurous acid electrolyzers, a key component of the Sulfur Cycle, are used in new generation combined sulfuric acid/hydrogen plants in the late 1980's, the full Sulfur Cycle is deployed in Solar Chemical plants in the 1990's, and the era of large scale merchant hydrogen production, with very high temperature nuclear reactors providing the energy needs of the Sulfur Cycle, starting after the turn of the century. This scenario can come to pass by diligently following the program path upon which we are now embarked.

Literature Cited

1. Farbman, G. H.; <u>NASA CR-134976</u>; "The Conceptual Design of an Integrated Nuclear Hydrogen Production Plan Using the Sulfur Cycle Water Decomposition System"; Westinghouse Electric Corporation; <u>April 1976</u>.

2. Farbman, G. H.; <u>FE-2262-15</u>; "Hydrogen Generation Process - Final Report"; <u>June 1977</u>.

3. Farbman, G. H.; Krasicki, B. R.; Hardman, C. C.; Lin, S. S.; Parker, G. H.; <u>EPRI-EM-789</u>; "Economic Comparison of Hydrogen Production Using Sulfuric Acid Electrolysis and Sulfur Cycle Water Decomposition"; <u>March 1978</u>; prepared by Westinghouse AESD for Electric Power Research Institute; Palo Alto, California 94304.

RECEIVED July 12, 1979.

APPENDIX

ECONOMIC BASIS

USE MIDYEAR 1979 DOLLARS   BASIS

    Project Life   20 years
    Operating Factor   330 days per year

Capital Investment

    Cost of capital   10%
    Working capital   60 day inventory
                                                   60 day cash supply
    Land required $5,000 per acre
    Startup expense and organization   2% of capital invest-
                                                       ment

Annual Operating Cost

    Feedstock, $/ton or $/MM Btu or ¢/lb
    Utilities
        Power ($/kwh)   0.030
        Water ($/Mgal)   0.05
        Fuel ($/MM Btu)   2.40 (delivered
                                               3.00 (no. 2 fuel)
    Operating labor $18,000/manyear
    Operating labor supervision   15% of operating labor
    Maintenance
        Labor   2% of facilities invest-
                                                   ment
        Supervision   15% of maintenance labor
        Materials   2% of facilities invest-
                                                   ment
    Administrative and support labor   20% of all other labor
    Payroll extras (fringe benefits,
        etc.)   20% of all other labor
    Insurance   2% of facilities invest-
                                                   ment
    General Administrative expenses   2% of facilities invest-
                                                   ment
    Taxes (local, state & federal)   50% of net profit
      (No investment tax credit)
    Depreciation   Straight line
    Depletion allowance   Specify method used
    By-product credit (if applicable)
      List quantity of each as $/ton or $/MM Btu or ¢/lb

INDEX

INDEX